Advances of Ceramic
and Alloy Coatings

Advances of Ceramic and Alloy Coatings

Editors

Yi Yang
Xinming Wang
Fucheng Yin

 Basel • Beijing • Wuhan • Barcelona • Belgrade • Novi Sad • Cluj • Manchester

Editors
Yi Yang
School of Materials Science
and Engineering
Xiangtan University
Xiangtan
China

Xinming Wang
School of Materials Science
and Engineering
University
Xiangtan
China

Fucheng Yin
School of Materials Science
and Engineering
Xiangtan University
Xiangtan
China

Editorial Office
MDPI
St. Alban-Anlage 66
4052 Basel, Switzerland

This is a reprint of articles from the Special Issue published online in the open access journal *Coatings* (ISSN 2079-6412) (available at: www.mdpi.com/journal/coatings/special_issues/ceramic_alloy).

For citation purposes, cite each article independently as indicated on the article page online and as indicated below:

Lastname, A.A.; Lastname, B.B. Article Title. *Journal Name* **Year**, *Volume Number*, Page Range.

ISBN 978-3-7258-1048-2 (Hbk)
ISBN 978-3-7258-1047-5 (PDF)
doi.org/10.3390/books978-3-7258-1047-5

© 2024 by the authors. Articles in this book are Open Access and distributed under the Creative Commons Attribution (CC BY) license. The book as a whole is distributed by MDPI under the terms and conditions of the Creative Commons Attribution-NonCommercial-NoDerivs (CC BY-NC-ND) license.

Contents

About the Editors . . . . **vii**

Zhiqiang Zhou, Dajun Duan, Shulan Li, Deen Sun, Jiahui Yong and Yongbing Jiang et al.
Microstructure and High-Temperature Properties of Cr_3C_2-25NiCr Nanoceramic Coatings Prepared by HVAF
Reprinted from: *Coatings* **2023**, *13*, 1741, doi:10.3390/coatings13101741 . . . . **1**

Bing Hu, Wei Xie, Wenhui Zhong, Dan Zhang, Xinming Wang and Jingxian Hu et al.
The Effect of Pulling Speed on the Structure and Properties of DZ22B Superalloy Blades
Reprinted from: *Coatings* **2023**, *13*, 1225, doi:10.3390/coatings13071225 . . . . **16**

Shaohua Zhang, Jinfang Wang, Meng Zhang, Longqi Ding, Huijun Chan and Xiyu Liu et al.
Mechanical and Tribological Behaviors of Hot-Pressed SiC/SiC_w-Y_2O_3 Ceramics with Different Y_2O_3 Contents
Reprinted from: *Coatings* **2023**, *13*, 956, doi:10.3390/coatings13050956 . . . . **33**

Jiansheng Li, Zuyuan Xu, Yu Zhao, Wei Jiang, Wenbo Qin and Qingzhong Mao et al.
Interfacial Microstructure and Shear Behavior of the Copper/Q235 Steel/Copper Block Fabricated by Explosive Welding
Reprinted from: *Coatings* **2023**, *13*, 600, doi:10.3390/coatings13030600 . . . . **46**

Jingqiang Zhang, Penghui Yang and Rong Wang
Investigation of the Microstructures and Properties of B-Bearing High-Speed Alloy Steel
Reprinted from: *Coatings* **2022**, *12*, 1650, doi:10.3390/coatings12111650 . . . . **55**

Longjie Lai, Heng Wu, Guobing Mao, Zhengdao Li, Li Zhang and Qi Liu
Microstructure and Corrosion Resistance of Two-Dimensional TiO_2/MoS_2 Hydrophobic Coating on AZ31B Magnesium Alloy
Reprinted from: *Coatings* **2022**, *12*, 1488, doi:10.3390/coatings12101488 . . . . **68**

Tong Liu, Jiansheng Li, Kezhang Liu, Mengmeng Wang, Yu Zhao and Zhongchen Zhou et al.
Effect of the Testing Temperature on the Impact Property of a Multilayered Soft–Hard Copper–Brass Block
Reprinted from: *Coatings* **2022**, *12*, 1236, doi:10.3390/coatings12091236 . . . . **80**

Jia Lou, Beibei Ren, Jie Zhang, Hao He, Zonglong Gao and Wei Xu
Evaluation of Biocompatibility of 316 L Stainless Steels Coated with TiN, TiCN, and Ti-DLC Films
Reprinted from: *Coatings* **2022**, *12*, 1073, doi:10.3390/coatings12081073 . . . . **90**

Rong Wu, Qionghua Yi, Shan Lei, Yilong Dai and Jianguo Lin
Design of Ti-Zr-Ta Alloys with Low Elastic Modulus Reinforced by Spinodal Decomposition
Reprinted from: *Coatings* **2022**, *12*, 756, doi:10.3390/coatings12060756 . . . . **102**

Zihan Wang, Faguo Li, Xiaoyuan Hu, Wei He, Zhan Liu and Yao Tan
Preparation of Ti-Al-Si Gradient Coating Based on Silicon Concentration Gradient and Added-Ce
Reprinted from: *Coatings* **2022**, *12*, 683, doi:10.3390/coatings12050683 . . . . **109**

Yongming Yan, Zixiang Luo, Ke Liu, Chen Zhang, Maoqiu Wang and Xinming Wang
Effect of Cryogenic Treatment on the Microstructure and Wear Resistance of 17Cr2Ni2MoVNb Carburizing Gear Steel
Reprinted from: *Coatings* **2022**, *12*, 281, doi:10.3390/coatings12020281 . . . . **120**

Wenhui Xiao, Ying Yang, Zhipeng Pi and Fan Zhang
Phase Stability and Mechanical Properties of the Monoclinic, Monoclinic-Prime and Tetragonal REMO$_4$ (M = Ta, Nb) from First-Principles Calculations
Reprinted from: *Coatings* **2022**, *12*, 73, doi:10.3390/coatings12010073 . . . . **133**

Jia Lou, Zonglong Gao, Jie Zhang, Hao He and Xinming Wang
Comparative Investigation on Corrosion Resistance of Stainless Steels Coated with Titanium Nitride, Nitrogen Titanium Carbide and Titanium-Diamond-like Carbon Films
Reprinted from: *Coatings* **2021**, *11*, 1543, doi:10.3390/coatings11121543 . . . . **148**

About the Editors

Yi Yang

Yi Yang is an associate professor and doctoral supervisor at Xiangtan University. He is mainly engaged in the design and preparation of high-entropy alloys, high-temperature corrosion mechanisms of high-entropy alloys, high-temperature corrosion studies of high-entropy ceramics and coating materials, and also has a strong research interest in transmission electron microscopy methods, such as high-resolution image simulation of TEM, high-angle annular dark-field image simulation, and space-group determination methods of crystalline material.

Xinming Wang

Xinming Wang is a researcher at the Key Laboratory of Materials Design and Preparation Technology of Hunan Province, Xiangtan University. His research areas include alloy thermodynamics, additive manufacturing, and material forming simulation. His publication list includes approximately 40 papers in peer-reviewed journals.

Fucheng Yin

Fucheng Yin is a professor and a doctoral supervisor at Xiangtan University. He is a member of the Chinese Phase Diagram Committee, a director of the Chinese Materials Research Society, and a director of the Materials Division of the Chinese Mechanical Engineering Society. He graduated from Central South University in 1986 with a B.S. degree in Metal Physics, received his M.S. degree in Metal Materials and Heat Treatment in 1989 from the Chinese Academy of Sciences (CAS), received his Ph.D. degree in Materials Physics and Chemistry in 2004, and did his post-doctoral work from September 2005 to September 2006 at the University of Montpellier, France. He is mainly engaged in phase diagram determination and thermodynamic calculation, material surface modification and design of high-temperature alloys and corrosion-resistant alloys, etc. He has published more than 80 papers in international journals and 1 book; 16 national patents; and 5 provincial and ministerial awards. He was awarded a total of 16 patents for his inventions.

Article

Microstructure and High-Temperature Properties of Cr_3C_2-25NiCr Nanoceramic Coatings Prepared by HVAF

Zhiqiang Zhou [1], Dajun Duan [1,*], Shulan Li [1], Deen Sun [2], Jiahui Yong [1], Yongbing Jiang [1], Wu He [1] and Jian Xu [1]

1. Chongqing Chuanyi Control Valve Co., Ltd., Chongqing 400707, China; zhouzzq6@gmail.com (Z.Z.)
2. School of Materials and Energy, Southwest University, Chongqing 400715, China; deen_sun@cqu.edu.cn
* Correspondence: ddj@siccv.com

Abstract: The study examines the microstructure and high-temperature properties of Cr_3C_2-25NiCr nanoceramic coatings on 316H high-temperature-resistant stainless steel that were prepared by high-velocity air–fuel spraying (HVAF) technology. The micromorphology, phase composition, fracture toughness, high-temperature hardness, high-temperature friction, and wear properties of the coating were studied by scanning electron microscopy (SEM), X-ray diffraction (XRD), high-temperature Vickers hardness tester, high-temperature friction and wear tester, and surface profiler. The results show that the Cr_3C_2-25NiCr coating prepared by HVAF presents a typical thermal spraying coating structure, with a dense structure and a porosity of only 0.34%. The coating consists of a Cr_3C_2 hard phase, a NiCr bonding phase, and a small amount of Cr_7C_3 phase; The average microhardness of the coating at room temperature is 998.8 $HV_{0.3}$, which is about five times higher than that of 316H substrate. The Weibull distribution of the coating is unimodal, showing stable mechanical properties. The average microhardness values of the coating at 450 °C, 550 °C, 650 °C, and 750 °C are 840 $HV_{0.3}$, 811 $HV_{0.3}$, 729 $HV_{0.3}$, and 696 $HV_{0.3}$ respectively. The average friction coefficient of the Cr_3C_2-25NiCr coating initially decreases and then increases with temperature. During high-temperature friction and wear, a dark gray oxide film forms on the coating surface. The formation speed of the oxide film accelerates with increasing temperature, shortening the running-in period of the coating. The oxide film acts as a lubricant, reducing the friction coefficient of the coating. The Cr_3C_2-25NiCr coating exhibits exceptional high-temperature friction and wear resistance, primarily through oxidative wear. The Cr_3C_2-25NiCr coating exhibits outstanding high-temperature friction and wear resistance, with oxidative wear being the primary wear mechanism at elevated temperatures.

Keywords: HVAF; Cr_3C_2-25NiCr coating; micromorphology; mechanical properties; high-temperature friction and wear properties

1. Introduction

316H stainless steel is frequently employed as a high-temperature-resistant material for valve stem components in high-temperature, high-pressure steam conditions, with a maximum operating temperature reaching 816 °C. Nevertheless, at elevated temperatures, the hardness of 316H stainless steel is relatively low, rendering it susceptible to erosion caused by gas or particulate-laden gas-solid two-phase media. Consequently, there is a need to elevate the surface hardness of 316H stainless steel through surface-hardening techniques.

Cr_3C_2-25NiCr currently stands as one of the most extensively employed nanoceramic composite materials, among which the NiCr alloy component of this composite boasts commendable resistance to heat-induced corrosion and exceptional capabilities in countering high-temperature oxidation. The elevated hardness attributed to Cr_3C_2 further enhances its exceptional resistance to high-temperature friction and wear [1]. Consequently, the Cr_3C_2-25NiCr coating exhibits commendable performance in withstanding elevated temperatures, displaying resilience against high-temperature erosion, oxidation, and wear at 900 °C. This prowess has propelled its successful integration across industries encompassing petroleum refining, thermal power generation, aerospace, metallurgical machinery, and other fields [2,3]. Notably, thermal spray technology emerges as a pivotal technique in the fabrication of Cr_3C_2-25NiCr coatings [4].

The utilization of high-velocity oxygen fuel spraying (HVOF) technology is prevalent in the preparation of Cr_3C_2-25NiCr coatings. However, the HVOF technology uses oxygen as the combustion-assisting gas, and the metal-powder particles are in a rich oxygen atmosphere during the spraying process, which is prone to thermal decomposition of powder oxidation or carbides [5]. High-velocity air–fuel spraying (HVAF) technology is a new technology developed in recent years. HVAF uses compressed air instead of expensive oxygen as the combustion-assisting gas and adopts a gas-cooling method. HVAF has a higher spraying flame velocity and lower flame temperature than HVOF technology, which helps to form metal coatings with high density, low oxide content, and high bonding strength [6].

Mathiyalagan et al. [7] utilized the HVAF technique to fabricate Ni-P coatings containing c-BN on the surface of 350LA, resulting in a hardness enhancement of 47% and a reduction of the wear rate by two orders of magnitude. The wear resistance was significantly improved. The dimensions of the combustion chamber also exert an influence on the wear resistance of the coating. The research findings indicate that a relatively larger combustion chamber can notably diminish the quantity of nondeformed particles within the coating, thereby leading to lower porosity and relatively higher hardness within the coating, thus achieving the objective of enhancing the coating's wear resistance [8]. Alroy [9] investigated the influence of process parameters on the corrosion performance of HVAF-prepared Cr_3C_2-25NiCr coatings. It was observed that by using fine-grained powders and a medium-length nozzle during the coating preparation, the coating exhibited reduced porosity and higher density, leading to a corrosion rate decrease of 40%–45% compared to the substrate. Furthermore, the spray flame velocity and flame temperature of HVAF were reported to be 700–1200 m/s and 1800 °C, respectively [10,11]. The semimolten sprayed powders have a very short flight time in the air, which can effectively suppress oxidation, decomposition, and decarburization of the powder so that the majority of hard phases can be retained. Therefore, HVAF technology has received a lot of attention [12,13].

In electric power, metallurgy, and other industries, various ball valves are installed on pipelines as switching control equipment for fluid media. The temperature of high-temperature heat transfer oil, high-temperature flue gas, high-temperature steam, and other media in the pipeline is often as high as about 600 °C, and its key parts are easily damaged and fail, seriously affecting the normal operation of the ball valve. Therefore, the surface strengthening of the critical components is very important. Usually, researchers will spray a hardened layer on the ball core, valve seat, and other seals of the ball valve to ensure the high-temperature oxidation resistance and wear resistance of the valve under high-temperature conditions. However, at present, there is limited research on the high-temperature performance of Cr_3C_2-25NiCr nanoceramic coatings prepared using HVAF technology. Currently, there is limited research on the high-temperature performance of Cr_3C_2-25NiCr nanoceramic coatings prepared using HVAF technology. In order to enhance the high-temperature surface-wear resistance of AISI 316H stainless steel, based on HVAF technology and optimized process parameters, Cr_3C_2-25NiCr nanoceramic coatings were prepared on the surface of a 316H stainless steel substrate. Through a series of experiments, the microstructural morphology, phase composition, high-temperature hardness, and high-temperature friction-wear performance of the coating were investigated. Additionally, the high-temperature friction-wear mechanism was explored, thereby establishing the feasibility of applying the Cr_3C_2-25NiCr nanoceramic coating prepared by HVAF technology under high-temperature operating conditions.

2. Materials and Methods

2.1. Coating Preparation

The substrate material for the coating preparation is 316H stainless steel; its chemical composition is shown in Table 1. The chemical composition of the substrate was determined using ICAP 7000 inductively coupled plasma optical emission spectrometry (ICP-OES, Thermo Fisher Scientific, Waltham, MA, USA) and a CS-988A carbon sulfur analyzer (CSA, Wuxi Tianmu Instrument Technology Co., Ltd., Wuxi, China). The specimen size was 30 mm × 15 mm × 8 mm. To remove any impurities, such as oxide or oil residues that may remain on the surface of the sample, the 316H substrate was repeatedly cleaned using acetone and absolute ethanol and dried with compressed air. After cleaning, the substrate was sandblasted to ensure optimal surface roughness, thereby enhancing the bond strength between the coating and the substrate.

Table 1. Chemical composition of 316H stainless steel (wt.%).

C	Mn	Si	S	P	Ni	Cr	Mo	Fe
0.07	1.83	0.86	0.02	0.03	13.25	17.15	2.31	Bal.

The spraying powder used for the experiments is a Cr_3C_2-25NiCr powder produced by Luoyang Jinglu Hard Alloy Tool Co., Ltd. (Luoyang, China). The powder is manufactured using a method involving Ni and Cr encapsulation of Cr_3C_2, and its composition is detailed in Table 2. Powder particle-size distribution is measured using a laser particle-size analyzer (Microtrac S3500, Boca Raton, FL, USA).

Table 2. Chemical composition of Cr_3C_2-25NiCr powder (wt.%).

C	Ni	Fe	Cr
10.66	20.35	0.41	Bal.

The coating preparation is conducted using the M2 HVAF (UNIQUECOAT, Oilville, VA, USA) supersonic flame-spraying system. In this experiment, the application employs optimized spray-process parameters for batch production (as presented in Table 3) targeting the fabrication of Cr_3C_2-25NiCr nanoceramic coatings on 316H stainless steel.

Table 3. Process parameters of HVOF and HVAF.

Air pressure (MPa)	Propane pressure (MPa)	Nitrogen flow (L/min)	Airflow (m^3/min)	Spraying distance (mm)	Powder speed (mm/s)	Powder feeding (g/min)
0.54	0.49	60	20	230	800	110

Coating cross-section specimens are prepared using a wire-cutting process. Following thermal embedding, coarse grinding, fine grinding, and polishing, the surface roughness of the coating is achieved at Ra < 0.2 μm.

2.2. Performance Characterization

The microstructure of Cr_3C_2-25NiCr powder and coatings was observed using the Navo Nano SEM 450 field emission scanning electron microscope (SEM, FEI, Hillsboro, OR, USA). Additionally, the chemical composition of the coating was analyzed by Quantax-200 EDS (Bruker, Billerica, MA, USA) spectrometry.

The coating cross-section microstructure is examined using the Axio Observer 3 m (Carl Zeiss AG, Oberkochen, Germany) research-grade metallographic microscope. The average porosity of 5 different fields is calculated using the Pro Imaging metallographic intelligent analysis system.

The phase composition of Cr_3C_2-25NiCr powder and the coatings is analyzed using the Rigaku D/MAX 2500 PC X-ray diffractometer (XRD, Rigaku, Tokyo, Japan). The samples are subjected to phase testing with a tube voltage of 30 kV, tube current of 20 mA, scanning angle range of 20°–90°, and continuous scanning rate of 0.03 °/s.

Microhardness of the coatings was measured using the INNOVATEST FALCON 500 (Eindhoven, The Netherlands) Vickers hardness tester with a load of 300 g and a loading time of 15 s. Additionally, 15 hardness values are obtained in the field-of-view area of the coating cross-section, and the Weibull distribution method is employed to explore the hardness distribution characteristics of the coating at room temperature. Equations (1) and (2) were employed to characterize the hardness-distribution characteristics of the coating [14].

In the equation, where $F(x)$ represents the hardness probability distribution function of the coating, m stands for the modulus of the Weibull distribution function, reflecting the degree of discreteness of numerical expressions. The parameter x represents the measured hardness value, while x_0 signifies the hardness value obtained after sorting the hardness points in ascending order, accounting for 63.2% of the total hardness points.

$$\ln\{\ln[\frac{1}{1-F(x)}]\} = m[\ln(x) + \ln(x_0)] \tag{1}$$

$$F(x) = \frac{i - 0.5}{n} \tag{2}$$

In the equation, where n represents the total number of hardness indentations, and i indicates the sequence number of hardness values arranged in ascending order.

The high-temperature microhardness of the coatings is evaluated using the HVZHT-30 high-temperature vickers hardness tester (ZONE-DE, Shandong, China), applying a load of 500 g and a dwell time of 10 s. The holding time at each temperature is 10 min. Three hardness values are recorded at each temperature, and real-time high-speed images of the indentations are captured.

The UMT-3 multifunctional high-temperature friction and wear tester (CETR, San Jose, CA, USA) is utilized to assess the friction-wear performance of the Cr_3C_2-25NiCr coating in high-temperature air at temperatures of 450 °C, 550 °C, 650 °C, and 750 °C, respectively. The test is pin–disc contact, and for every single test, a test duration of 30 min and a new Al_2O_3 (the Mohs hardness scale is 9) ball with a diameter of 9.5 mm were used, with a load of 5 N and a disc rotation speed of 150 r/min.

The DektakXT (Bruker, Billerica, MA, USA) probe surface profiler measures the cross-sectional profiles of the wear scars.

3. Results

3.1. Powder Morphology

The microstructure of the agglomerated Cr_3C_2-25NiCr spherical nanopowder for supersonic spray prepared through the gas atomization (agglomerating sintering) process is shown in Figure 1a. It is evident that the powder possesses a high degree of sphericity and excellent dispersion, indicating a favorable flowability of the powder.

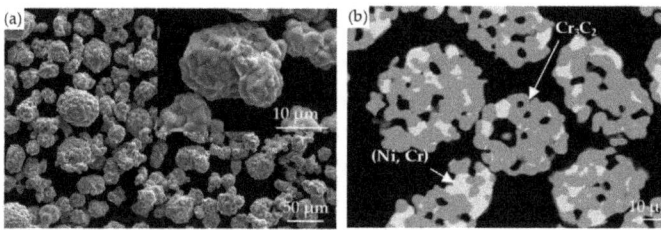

Figure 1. BSE image of Cr_3C_2-25NiCr powder (**a**) surface morphology; (**b**) cross-sectional morphology.

Figure 1b shows the powder cross-sectional metallurgical structure. The metallurgical structure within the cross-section reveals the Cr_3C_2 phase wetted by the NiCr alloy γ phase [15]. The gray region corresponds to the Cr_3C_2 phase, while the silver-white region represents the NiCr alloy γ phase. It can be observed that the Cr_3C_2 particles in the powder were sintered, and the NiCr alloy bonding phase uniformly filled the interstices between the hard particles, forming a stable skeleton-network structure.

Figure 2 illustrates the particle-size distribution of the powder, conforming to a standard Gaussian distribution. Among the parameters, dmean = 32 μm, d10 = 18 μm, and d90 = 52 μm.

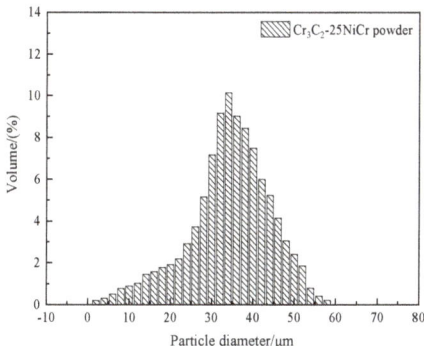

Figure 2. Particle-size distribution of Cr_3C_2-25NiCr powder.

3.2. XRD Phase Composition

Figure 3 exhibits the X-ray diffraction (XRD) spectra of Cr_3C_2-25NiCr coatings prepared by HVAF, as well as the initial powder. It is evident that the sprayed coating retains the crystalline phases present in the raw powder. The Cr_3C_2-25NiCr powder and coating are primarily composed of the Cr_3C_2 hard phase and NiCr bonding phase. Additionally, a small amount of the Cr_7C_3 diffraction peak is observed in the coating. This can be attributed to the partial decarburization of Cr_3C_2 during the high-temperature flame in the spraying process [16]. Both Cr_3C_2 and Cr_7C_3 possess high hardness and high melting points, which contribute to the wear resistance and high-temperature hardness of the coating [3].

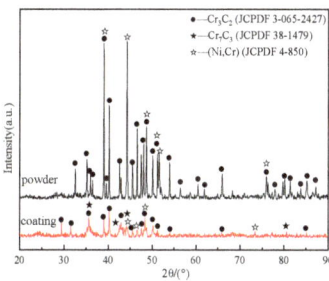

Figure 3. XRD pattern of Cr_3C_2-25NiCr powder and coating.

Notably, there is a substantial difference in the intensity of diffraction peaks between the powder and the coating. The diffraction peaks of the coating exhibit a significant broadening phenomenon, indicating the generation of an amorphous phase during the transformation from powder to coating. This can be attributed to the rapid cooling of molten or semimolten droplets upon reaching the substrate during the supersonic spray process. This rapid cooling suppresses crystal growth, resulting in a disordered accumulation of solidifications and an amorphous coating appearance [17]. Additionally, research suggests that the formation of the amorphous phase during HVAF spraying of Cr_3C_2-25NiCr coatings is related to the severe plastic deformation occurring upon the high-velocity impact of powder particles on the substrate [18]. Consequently, in the process of preparing ceramic coatings through supersonic flame spraying, the coating consists of both crystalline and amorphous phases, with the presence of the amorphous phase potentially enhancing wear resistance to a certain extent [19].

3.3. Section Morphology

Figure 4 shows the SEM cross-sectional morphology of Cr_3C_2-25NiCr nanoceramic coatings prepared by HVAF. From Figure 4a, it can be observed that the thickness of the Cr_3C_2-25NiCr coating is approximately 260 μm, forming a mechanical connection with the substrate through mechanical interlocking. The coating exhibits density without visible cracks or significant defects. During HVAF spraying, with a flame temperature of 1800 °C, the NiCr alloy, due to its lower melting point, rapidly melts and wets the surrounding ultrafine Cr_3C_2 nanoparticles, leading to deformation and melting of the outer layers of the Cr_3C_2 particles. These molten particles impact the substrate at high speed, flattening and forming a typical thermal spray layer structure. Simultaneously, a small fraction of Cr_3C_2 undergoes decarburization, with carbon atoms diffusing into the molten NiCr phase to form a solid solution [20,21]. The coating microstructure is characterized by a continuous and well-melted NiCr bonded phase uniformly dispersed with carbide hard particles, such as Cr_7C_3, among which the light gray phase corresponds to the NiCr bonded phase, while the dark gray phase represents the hard particles.

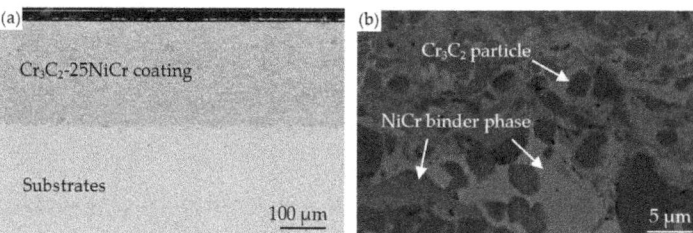

Figure 4. BSE cross-section micrographs of Cr_3C_2-25NiCr coating: (**a**) lower magnification; (**b**) higher magnification.

Porosity is one of the significant parameters of coating performance, which will significantly affect the microhardness, wear resistance, and corrosion resistance of the coating. The pores of the coating are mainly distributed at the boundaries of hard particles, which is primarily due to insufficient deformation of powder particles and incomplete overlap during deposition. Another portion of the pores is caused by the solidification shrinkage of the fully melted binder phase that cannot be compensated. Employing the "gray level method", the calculated porosity of the coating is 0.34%, indicating a high hardness value [22].

Figure 5 displays the as-sprayed surface morphology of the Cr_3C_2-25NiCr coating. It is apparent that the surface of the coating is primarily composed of a multitude of unmelted powder particles along with a small fraction of the completely melted solidification area. This is due to the fact that the HVAF spraying technique employs propane–air as fuel. Compared to HVOF supersonic flame-spraying technology, which uses aviation kerosene as fuel, the heat during the spraying process is low. As a result, some powder particles find it challenging to attain their melting points, and molten or semimolten particles in the flame flow are in a solid–liquid mixed state before impacting the substrate. In the case of HVAF technology, the high powder-injection velocity leads to the flattening of most unmelted or semimelted powder particles. These particles are then layered and bonded onto the substrate surface through the application of significant impact forces and plastic deformation pressures [23].

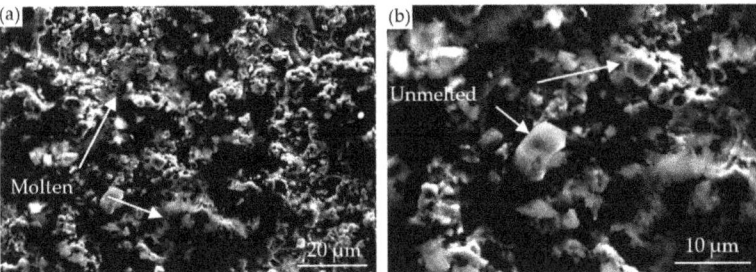

Figure 5. As-spraying surface morphology of Cr_3C_2-25NiCr coating: (**a**) low magnification; (**b**) high magnification.

3.4. Microhardness and Weibull Distribution

Figure 6 depicts the microhardness distribution across the cross-section of the Cr_3C_2-25NiCr nanoceramic coating. The average microhardness of the F316H substrate is approximately 210 $HV_{0.3}$, and the microhardness of the substrate near the coating interface tends to increase slightly, reaching around 400 $HV_{0.3}$. This increase can be attributed to the "shot peening effect" induced by the powerful impact of the powder particles on the substrate during a high-speed deposition process. Therefore, the effect of work hardening is produced [24]. The result shows that the closer to the coating interface, the higher the microhardness of the substrate. And at the coating-substrate interface, the microhardness reached 750 $HV_{0.3}$. As can be seen from Figure 6, with the increase in coating thickness, the hardness value of the coating increases slightly and then decreases. It spans from 921 $HV_{0.3}$ to 1090 $HV_{0.3}$, and the average microhardness is about 998 $HV_{0.3}$, which is about five times higher than that of the F316H substrate

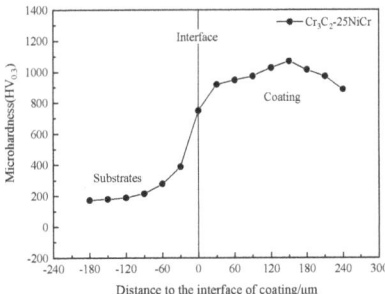

Figure 6. Microhardness profile of Cr_3C_2-25NiCr coatings.

The Weibull distribution parameters and the Weibull distribution curve of microhardness of the Cr_3C_2-25NiCr coating at room temperature are shown in Table 4 and Figure 7, respectively. The distribution of Weibull hardness points of the coating aligns well with the linear fit and demonstrates a unimodal distribution characteristic [25]. The microhardness values of the coating exhibit a narrow distribution range and small hardness extreme values, signifying high uniformity and stable mechanical performance.

Table 4. Weibull distribution parameters of microhardness of Cr_3C_2-25NiCr coating.

Hardness(x) (HV$_{0.3}$)	ln(x)	m	n	Average hardness (HV$_{0.3}$)
921~1090	6.83~6.99	24.45	15	998

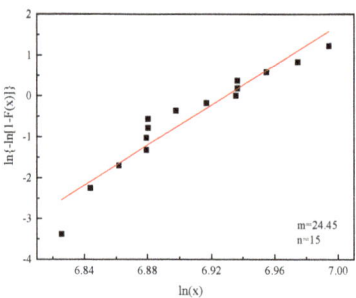

Figure 7. Weibull distribution of microhardness for Cr_3C_2-25NiCr coating.

3.5. High-Temperature Hardness

As indicated in Figure 8, the average microhardness of the Cr_3C_2-25NiCr coating at room temperature, 450 °C, 550 °C, 650 °C, and 750 °C are 998 HV$_{0.3}$, 840 HV$_{0.3}$, 811 HV$_{0.3}$, 729 HV$_{0.3}$, and 696 HV$_{0.3}$, respectively. With the increase in temperature, the hardness of the coating decreases. This phenomenon can be attributed to the reduction in both the grain and the grain boundary strength of the material as temperature rises, leading to a high-temperature softening of the coating [26]. Remarkably, even with a temperature increase from 450 °C to 750 °C, the high-temperature hardness of the Cr_3C_2-25NiCr coating only experiences a decrease of approximately 140 HV$_{0.3}$, and it still has a high high-temperature hardness.

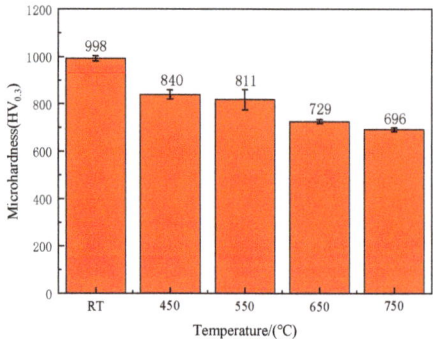

Figure 8. Hardness of Cr_3C_2-25NiCr coatings at high temperature.

Figure 9 illustrates the real-time microhardness indentation morphology of the Cr_3C_2-25NiCr coating at elevated temperatures of 450 °C, 550 °C, 650 °C, and 750 °C. Notably, the indentation morphologies and sizes appear relatively consistent across the temperatures, and no cracks are observed around the diagonals of the indentations. This observation indicates that the Cr_3C_2-25NiCr coating maintains a high level of high-temperature hardness and fracture toughness.

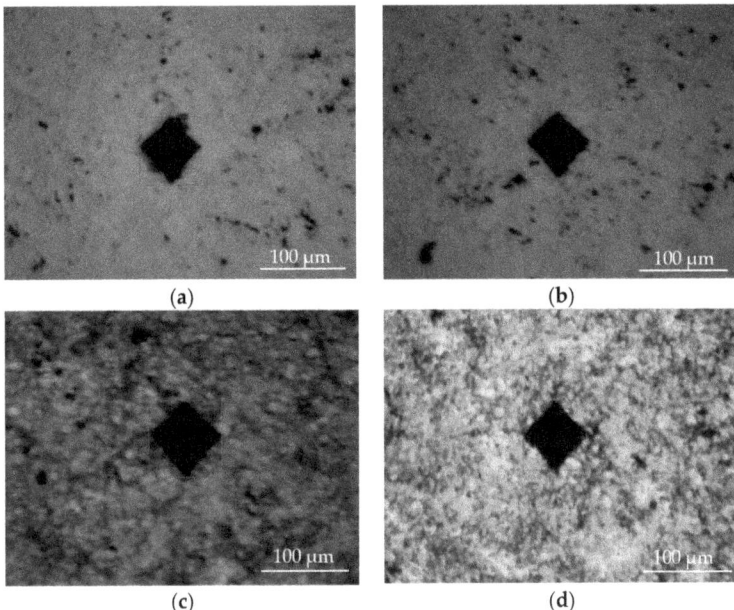

Figure 9. Indentation Morphology of Cr_3C_2-25NiCr coating at high temperature: (**a**) 450 °C; (**b**) 550 °C; (**c**) 650 °C; (**d**) 750 °C.

With the increase in temperature, the contrast between the light gray Cr_3C_2 particles and the surrounding dark gray NiCr bonding phase becomes more distinct, rendering the morphology more discernible under optical microscopy. This phenomenon is attributed to the oxidation of Cr elements on the surface of the coating at high temperatures, resulting in the formation of an oxide film that significantly enhances the contrast between the hard phase and the bonding phase.

3.6. High-Temperature Friction and Wear

3.6.1. Coefficient of Friction

Figure 10 illustrates the friction coefficient curves and the average coefficient of friction for the Cr_3C_2-25NiCr coating at temperatures of 450 °C, 550 °C, 650 °C, and 750 °C. During the friction test, the friction coefficient of the coating experiences an initial rise followed by a decrease during the running-in period, after which it enters a stable period of small fluctuations. As the temperature increases, the running-in period gradually shortens. At 450 °C, 550 °C, and 650 °C, the running-in periods for the coating are 1056 s, 1010 s, and 987 s, respectively. At 750 °C, the running-in period is only 242 s and swiftly enters a stabilization period. With a rising temperature, the running-in period tends to shorten, and the friction and wear process rapidly enters a stable state.

Notably, at 450 °C, the Cr_3C_2-25NiCr coating exhibits the highest coefficient of friction, with an average value of 0.52 ± 0.02. At 550 °C, the lowest coefficient of friction is observed, with an average value of 0.44 ± 0.01. However, at 750 °C, the coefficient of friction increases to 0.48 ± 0.01. Moreover, at this temperature, the friction coefficient was the most stable, fluctuating only within a small range. And with the increase in sliding time, it shows an upward trend.

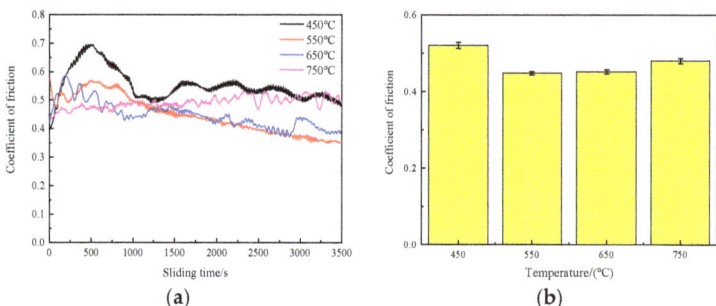

Figure 10. Friction coefficient curve and average friction coefficient of Cr_3C_2-25NiCr coatings at different temperatures: (**a**) friction coefficient curve; (**b**) mean friction coefficient.

The friction force of the material at high temperatures is mainly the plastic deformation of the surface and the role of the contact point between the friction pair and the surface of the material. With the increase of the test temperature, the plastic deformation force diminishes while the effect of contact points on the surface grows. At high temperatures, adhesive bonding occurs between the friction pair and the material at contact points on the surface. During sliding, the bonded points are sheared, and material transfer takes place at the interface. This alternating process of bonding and shearing causes minor oscillations in the friction curve [27,28]. After entering the stabilization period, the coefficient of friction stabilizes, indicating that the coating has not undergone wear failure. Consequently, the Cr_3C_2-25NiCr coating can play a certain protective role on the 316H substrate at a temperature of 450 °C–750 °C.

Figure 11 illustrates the corresponding cross-sectional profile curves and wear rates of the Cr_3C_2-25NiCr coating at temperatures of 450 °C, 550 °C, 650 °C, and 750 °C. It can be seen from the figure that, at 450 °C, the Cr_3C_2-25NiCr coating has the smallest abrasion area and the lowest wear rate, showing the best wear resistance. With an increasing temperature, the wear rate of the Cr_3C_2-25NiCr coating gradually rises, and the corresponding abrasion area gradually increases.

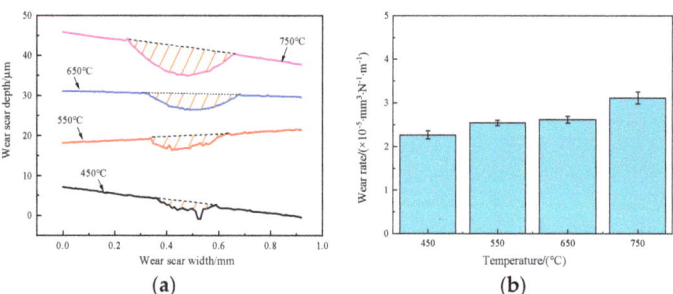

Figure 11. The wear scars profile of cross-section and wear rate of Cr_3C_2-25NiCr coatings at different temperatures: (**a**) wear-scars profile of cross-section; (**b**) wear rate.

At 450 °C, although the coating exhibits the highest friction coefficient (0.527), its hardness is also the highest (840 $HV_{0.3}$), resulting in a wear rate of $(2.16 \pm 0.03) \times 10^{-5}$ mm^3/(N·m). At 550 °C and 650 °C, the wear rates are $(2.52 \pm 0.01) \times 10^{-5}$ mm^3/(N·m) and $(2.68 \pm 0.01) \times 10^{-5}$ mm^3/(N·m), respectively, and the wear rates are very close, which is related to the minimal variation in its coefficient of friction, suggesting stable mechanical performance of the coating at these temperatures.

However, at 750 °C, the wear rate increases to $(2.97 \pm 0.02) \times 10^{-5}$ mm^3/(N·m). This increase is attributed to the higher tendency of softening in the NiCr bonding phase of the coating at this temperature [29], which reduces the cohesive strength of the coating. As a result, Cr_3C_2 (EDS result is shown in Figure 12h) undergoes secondary precipitation, leading to the detachment of hard particles [30]. The coating's shear resistance diminishes, resulting in the highest wear rate observed at this temperature.

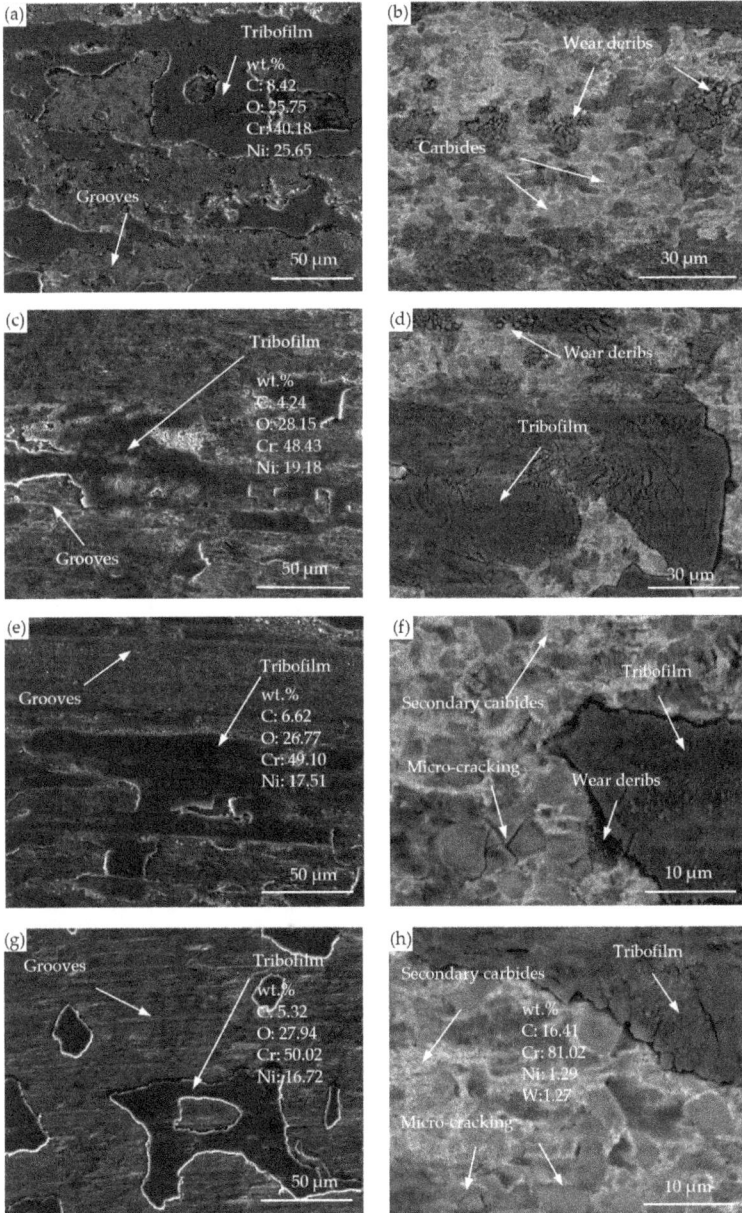

Figure 12. SEM micrographs of the wear scars of Cr_3C_2-25NiCr coating after wear testing: (**a,b**) low- and high-magnification picture at 450 °C; (**c,d**) low- and high-magnification picture at 550 °C; (**e,f**) low- and high-magnification picture at 650 °C; (**g,h**) low- and high-magnification picture at 750 °C.

3.6.2. Wear Mechanism

Figure 12 shows the wear-surface morphology of the Cr_3C_2-25NiCr coating at different temperatures. The surface exhibits distinctive furrows and extensive areas covered by flaky dark gray oxide film [31]. EDS analysis reveals that these oxide films primarily consist of the elements C, Cr, O, and Ni, of which the oxygen content is 25 wt.%–28 wt.%. This indicates that during high-temperature and oxidative friction and wear, a smooth oxide film (mainly Cr_2O_3 and CrO_3) is first formed on the surface of the Cr_3C_2-25NiCr coating [32].

These oxide-based friction films act as a nonplastic medium, existing between the coating surface and aluminum oxide balls, effectively preventing direct contact and reducing the friction coefficient. Additionally, the oxide-based friction films act as lubricants, further lowering the friction coefficient of the coating [33]. As a result, as shown in Figure 11, the coefficient of friction of the coating decreases when the temperature is 450 °C. With the increasing temperature, the running-in period of the coating shortens gradually. This is due to the accelerated formation rate of the oxide films containing metal Cr on the surface of the coating, and the oxide film plays a lubricating role between the friction pairs, leading to a rapid transition to the stable wear period.

Furthermore, it can be seen in Figure 12b,d that numerous fine debris particles are observed on the coating surface, which play a role in load distribution and also offer a protective effect; it will also have a protective effect and reduce the wear of the coating.

The friction and wear mechanism of the Cr_3C_2-25NiCr coating at high temperatures are shown in Figure 13. During high-temperature friction and wear, the formation and breakage of oxide films are in a dynamic equilibrium. In the process, the surface of the preferentially formed sheet of oxide film produces microcracks (as seen in Figure 12 f,h), and gradually loosens and detaches. Then, new fine debris is formed, and the newly exposed alloy area continues to be oxidized after the oxide film is peeled. The peeled debris recombines with the newly formed oxide film, creating a dynamic process of alternating oxidation and detachment. Therefore, at high temperatures, the primary wear mechanism of the Cr_3C_2-25NiCr coating is oxidative wear.

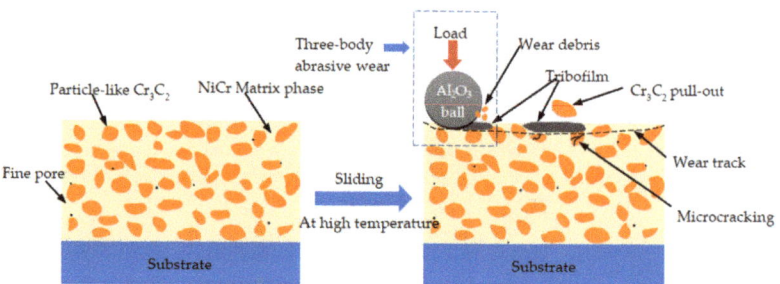

Figure 13. Schematic diagram of the wear mechanism of Cr_3C_2-25NiCr coating at high temperature.

In the high-magnification SEM images of Figure 12f,h, it can be observed that, at temperatures of 650 °C and 750 °C, some carbide hard particles undergo thermal and shear stresses, gradually developing microcracks under the fatigue action of cyclic load, which eventually causes the hard phase to peel off from the NiCr bonded phase and form pits. Additionally, it can be observed that, during prolonged high-temperature friction and wear processes, a secondary precipitation of the carbides occurs in the NiCr bonding phase [33], potentially leading to the formation of brittle regions and the shedding of carbide hard phases [30].

As the temperature increases from 450 °C to 750 °C, the oxide layer becomes thicker, giving rise to more debris and an oxide-based friction film. Along the edges of the oxide film in Figure 12d,f, there are a large number of fine oxide debris, which will be used as abrasive particles to form a three-body abrasive wear, resulting in a relatively high wear rate of the coating at 750 °C.

4. Conclusions

1. A Cr_3C_2-25NiCr nanoceramic coating was prepared using an air-assisted supersonic flame-spraying technique (HVAF) with a coating thickness of 260 μm. The coating exhibited a dense surface with low porosity (0.34%). The microstructure of the coating displayed a typical thermal spray-layer structure, consisting of a Cr_3C_2 hard phase, NiCr bonding phase, and a small amount of Cr_7C_3;
2. The average microhardness of the Cr_3C_2-25NiCr coating at room temperature was 998 $HV_{0.3}$, which is approximately five times higher than that of the 316H substrate. The Weibull distribution of hardness values for the coating showed a single peak feature, and small hardness extreme values, indicating stable mechanical performance;
3. The average microhardness values of the Cr_3C_2-25NiCr coating at 450 °C, 550 °C, 650 °C, and 750 °C were 840 $HV_{0.3}$, 811 $HV_{0.3}$, 729 $HV_{0.3}$, and 696 $HV_{0.3}$, respectively. With increasing temperature, the coating exhibited a decreasing trend in microhardness due to high-temperature softening. However, it still maintained relatively high hardness at elevated temperatures;
4. At 450 °C, the Cr_3C_2-25NiCr coating exhibits the best high-temperature friction and wear properties. The wear rates of the coating at 450 °C, 550 °C, 650 °C, and 750 °C were $(2.16 \pm 0.03) \times 10^{-5}$ mm^3/(N·m), $(2.52 \pm 0.01) \times 10^{-5}$ mm^3/(N·m), $(2.68 \pm 0.01) \times 10^{-5}$ mm^3/(N·m), and $(2.97 \pm 0.02) \times 10^{-5}$ mm^3/(N·m), respectively. With increasing temperature, the average friction coefficient of the Cr_3C_2-25NiCr coating shows a trend of initially decreasing and then increasing, corresponding to the gradual enlargement of the wear scar area;
5. During high-temperature friction and wear, a large area of sheet-like dark gray oxide film formed on the coating surface. As the temperature increased, the rate of oxide film formation accelerated, and the run-in period of the coating gradually shortened. This oxide film acted as a lubricant between the friction pairs, effectively reducing the friction coefficient of the coating at high temperatures;
6. The wear mechanism of the Cr_3C_2-25NiCr coating at high temperatures is mainly oxidative wear. At temperatures of 650 °C and 750 °C, certain carbide hard particles develop microcracks and undergo secondary precipitation, leading to the formation of a brittle zone. This zone, in conjunction with oxide debris, contributes to the occurrence of three-body abrasive wear.

Author Contributions: Writing—original draft, writing—review and editing, Z.Z.; Formal analysis, W.H.; Investigation, S.L.; Data curation, J.Y.; Writing—review and editing, D.S.; Visualization, Y.J.; software, J.X.; Project administration, funding acquisition, Supervision, D.D. All authors have read and agreed to the published version of the manuscript.

Funding: This research was funded by Major National Science and Technology Projects, grant number 2019ZX06002026-005, China. And the APC was funded by Chongqing Chuanyi Control Valve Co., Ltd., Chongqing, China.

Institutional Review Board Statement: Not applicable.

Informed Consent Statement: Not applicable.

Data Availability Statement: Not applicable.

Conflicts of Interest: The authors declare no conflict of interest.

References

1. Luo, X.; Wang, S.-Y.; Hong, Y.-K.; Wu, J.-Y.; Luo, W.-Q.; Zhong, Z.-Q.; Yang, Q.-M. Friction behavior of WC-12Co, WC-10Co4Cr and Cr_3C_2-25NiCr coatings prepared by HVOF in different media. *Rare Met. Mater. Eng.* **2022**, *51*, 4682–4688.
2. Chen, T.-Z.; Hu, H.-Y.; Liu, H.; Gao, M.-C.; Wu, Y. Preparation of modified Cr_3C_2-25NiCr coating by HVOF and its corrosion resistance analysis. *Mater. Prot.* **2020**, *53*, 46–48.

3. Alroy, R.J.; Kamaraj, M.; Sivakumar, G. Influence of processing condition and post-spray heat treatment on the tribological performance of high velocity air-fuel sprayed Cr_3C_2-25NiCr coatings. *Surf. Coat. Technol.* **2023**, *463*, 129498. [CrossRef]
4. Matikainen, V.; Peregrina, S.R.; Ojala, N.; Koivuluoto, H.; Schubert, J.; Houdková, Š.; Vuoristo, P. Erosion wear performance of WC-10Co4Cr and Cr3C2-25NiCr coatings sprayed with high-velocity thermal spray processes. *Surf. Coat. Technol.* **2019**, *370*, 196–212. [CrossRef]
5. Zhao, Z.-P.; Si, H.-L.; Li, Z.-S.; He, Q.-B.; Yi, T.-B.; Song, K.-Q.; Cong, D.-L. Microstructure and friction and wear properties of NiCr-Cr_3C_2 coating by APS and HVOF. *Surf. Technol.* **2019**, *48*, 225–230.
6. Wang, Y.-J.; Wu, Y.-P.; Duan, J.-Z.; Hong, S.; Cheng, J.-B.; Cheng, J.; Zhu, S.-S. Microstructure, corrosion resistance, and antibacterial property of HVAF-sprayed Cu55Ti25Zr15Ni5 coating. *J. Alloys Compd.* **2023**, *967*, 171705. [CrossRef]
7. Mathiyalagan, S.; Rossetti, M.; Björklund, S.; Senad, D.; Samuel, A.; Radek, M.; František, L.; Shrikant, J. High Velocity Air Fuel (HVAF) Spraying of Nickel Phosphorus-coated Cubic-boron Nitride Powders for Realizing High-performance Tribological Coatings. *J. Mater. Res. Technol.* **2022**, *18*, 59–74. [CrossRef]
8. Mahade, S.; Aranke, O.; Björklund, S.; Dizdar, S.; Awe, S.; Mušálek, R.; Lukac, F.; Joshi, S. Influence of processing conditions on the microstructure and sliding wear of a promising fe-based coating deposited by HVAF. *Surf. Coat. Technol.* **2021**, *409*, 126953. [CrossRef]
9. Alroy, R.J.; Pandey, R.; Kamaraj, M.; Sivakumar, G. Role of process parameters on microstructure, mechanical properties and erosion performance of HVAF sprayed Cr_3C_2-NiCr coatings. *Surf. Coat. Technol.* **2022**, *449*, 128941. [CrossRef]
10. Sadeghimeresht, E.; Hooshyar, H.; Markocsan, N. Oxidation behavior of HVAF-Sprayed NiCoCrAlY coating in H_2–H_2O environment. *Oxid. Met.* **2016**, *86*, 299–314. [CrossRef]
11. Sadeghimeresht, E.; Markocsan, N.; Nylén, P.; Björklund, S. Corrosion performance of bi-layer Ni/Cr_2C_3-NiCr HVAF thermal spray coating. *Appl. Surf. Sci.* **2016**, *369*, 470–481. [CrossRef]
12. Wang, Y.; Cui, Z.-C. The application and discussion of ac-hvaf spraying process in power station equipment. *Therm. Spray Technol.* **2011**, *3*, 44–47.
13. Vashishtha, N.; Sapate, S.G.; Sahariaha, B.J. Microstructural characterization and wear behaviour of high velocity oxy-fuel sprayed Cr_3C_2-25NiCr coating. *Mater. Today Proc.* **2018**, *5*, 17686–17693. [CrossRef]
14. Aw, P.K.; Tan, B.H. Study of microstructure, phase and microhardness distribution of HVOF sprayed multi-modal structured and conventional WC–17Co coatings. *J. Mater. Process. Technol.* **2006**, *174*, 305–311. [CrossRef]
15. He, J.H.; Lavernia, E.J. Precipitation phenomenon in nanostructured Cr_3C_2–NiCr coatings. *Mater. Sci. Eng. A* **2001**, *301*, 69–79. [CrossRef]
16. Chen, M. Structure and Wear Behavior of NiCr-Cr_3C_2 Coatings Prepared by High Velocity Oxy-Fuel Spray. *Mater. Prot.* **2016**, *49*, 23–26.
17. Wu, N.C.; Chen, K.; Sun, W.H.; Wang, J.Q. Correlation between particle size and porosity of fe-based amorphous coating. *Surf. Eng.* **2019**, *35*, 37–45. [CrossRef]
18. Bolelli, G.; Berger, L.M.; Börner, T.; Koivuluoto, H.; Matikainen, V.; Lusvarghi, L.; Lyphout, C.; Markocsan, N.; Nylén, P.; Sassatelli, P.; et al. Sliding and abrasive wear behaviour of HVOF- and HVAF-sprayed Cr_3C_2–NiCr hardmetal coatings. *Wear* **2016**, *358–359*, 32–50. [CrossRef]
19. Evans, A.G.; Wilshaw, T.R. Quasi-static solid particle damage in brittle solids—I. Observations analysis and implications. *Acta Metall.* **1976**, *24*, 939–956. [CrossRef]
20. Matthews, S.; Berger, L.M. Long-term compositional/microstructural development of Cr_3C_2-NiCr Coatings at 500 °C, 700 °C and 900 °C. *Int. J. Refract. Met. Hard Mater.* **2016**, *59*, 1–18. [CrossRef]
21. Geng, Z.; Li, S.; Duan, D.L.; Liu, Y. Wear behaviour of WC–Co HVOF coatings at different temperatures in air and argon. *Wear* **2015**, *330–331*, 348–353. [CrossRef]
22. Tang, Q.; Zhang, S.-D.; Xu, M.; Wang, J.-Q. Effect of ceramic particles on corrosion resistance of thermal sprayed stainless steel coating. *J. Mater. Eng.* **2021**, *49*, 125–135.
23. Wang, Y.Y.; Li, C.J.; Ohmori, A. Influence of substrate roughness on the bonding mechanisms of high velocity oxy-fuel sprayed coatings. *Thin Solid Film.* **2005**, *485*, 141–147. [CrossRef]
24. Zhang, X.-w.; Liu, G.; Yi, J.-c.; Liu, H.-x. TiN/Ti-Al-Nb composite coatings prepared by in-situ synthesis assisted laser cladding process on Ti6Al4V titanium alloy surface. *Surf. Technol.* **2020**, *49*, 61–68.
25. Wang, D.-f.; Ma, B.; Ma, L.-c.; Chen, D.-g.; Liu, H.-w.; Zhang, Y.-y.; Zhang, L.; Dai, Y.; Wu, J.-m.; Gao, F. Effect of wc grain size on the microstructure and mechanical properties of HVOF-sprayed WC-10Co4Cr coatings. *Powder Metall. Technol.* **2019**, *37*, 434–443.
26. Huang, X.-z.; Li, C.; Su, G.-c.; Lu, J.-c. A study on hi-temperature softening resistance of in-Situ $TiC_p/Fe,VC_p/Fe$ Composites. *Mod. Cast Iron* **2011**, *01*, 91–93.
27. Riahi, A.R.; Edrisy, A.; Alpas, A.T. Effect of magnesium content on the high temperature adhesion of Al–Mg alloys to steel surfaces. *Surf. Coat. Technol.* **2009**, *203*, 2030–2035. [CrossRef]
28. Jin, Y.X.; Lee, J.M.; Kang, S.B. Wear characteristics of dry sliding friction between the T6 treated A356/SiC composite and semi-metallic pad. *Rare Met. Mater. Eng.* **2008**, *37*, 2147–2151.
29. Li, H.; Cheng, X.-N.; Xie, C.-S.; Lu, P.C. Microstructure and the high temperature friction and wear behavior of Cr_3C_2-NiCr coating on CuCo2Be alloy by plasma spraying. *Rare Met. Mater. Eng.* **2014**, *43*, 2011–2016.

30. Chhabra, P.; Kaur, M.; Singh, S. High temperature tribological performance of atmospheric plasma sprayed Cr_3C_2-NiCr coating on H13 tool steel. *Mater. Today Proc.* **2020**, *33*, 1518–1530. [CrossRef]
31. Matikainen, V.; Bolelli, G.; Koivuluoto, H.; Sassatelli, P.; Lusvarghi, L.; Vuoristo, P. Sliding wear behaviour of HVOF and HVAF sprayed Cr_3C_2-based coatings. *Wear* **2017**, *388–389*, 57–71. [CrossRef]
32. Zhou, W.X.; Zhou, K.S.; Li, Y.X.; Deng, C.; Zeng, K. High temperature wear performance of HVOF-sprayed Cr_3C_2-WC-NiCoCrMo and Cr_3C_2-NiCr hardmetal coatings. *Appl. Surf. Sci.* **2017**, *416*, 33–44. [CrossRef]
33. Mahade, S.; Mulone, A.; Björklund, S.; Klement, U.; Joshi, S. Novel suspension route to incorporate graphene nano-platelets in HVAF-sprayed Cr_3C_2–NiCr coatings for superior wear performance. *J. Mater. Res. Technol.* **2021**, *13*, 498–512. [CrossRef]

Disclaimer/Publisher's Note: The statements, opinions and data contained in all publications are solely those of the individual author(s) and contributor(s) and not of MDPI and/or the editor(s). MDPI and/or the editor(s) disclaim responsibility for any injury to people or property resulting from any ideas, methods, instructions or products referred to in the content.

Article

The Effect of Pulling Speed on the Structure and Properties of DZ22B Superalloy Blades

Bing Hu [1], Wei Xie [1], Wenhui Zhong [1], Dan Zhang [2,3], Xinming Wang [2,3,*], Jingxian Hu [2,3], Yu Wu [2,3] and Yan Liu [2,3]

[1] Aecc South Industry Company Limited, Zhuzhou 412000, China; hb3455027@sina.com (B.H.); daydayup0920@sina.com (W.X.); zhongwenhui504@sina.com (W.Z.)
[2] Key Laboratory of Materials Design and Preparation Technology of Hunan Province, Xiangtan University, Xiangtan 411105, China; 202121551588@smail.xtu.edu.cn (D.Z.); jingxianhu@xtu.edu.cn (J.H.); wuyu@xtu.edu.cn (Y.W.); liuyan18773226132@163.com (Y.L.)
[3] School of Materials Science and Engineering, Xiangtan University, Xiangtan 411105, China
* Correspondence: wangxm@xtu.edu.cn; Tel.: +86-189-7522-8303

Abstract: DZ22B alloy is commonly used as a blade material for aircraft engines and gas turbines, and its preparation process is an important factor affecting its performance. In the present work, a reliable numerical model is established through ProCAST numerical simulation and auxiliary experimental verification methods, based on which the effect of casting speed on the grain and dendrite growth of DZ22B superalloy blades is studied. The results indicate that increasing the pulling speed can reduce the spacing between secondary dendrites, which is beneficial for the growth of dendrites. Based on numerical simulation and experimental verification, it is suggested that the pulling rate of the directional solidification DZ22B superalloy blade should be 6-2 mm/min variable speed pulling to improve the production success rate.

Keywords: directional solidification; DZ22B superalloy blades; ProCAST numerical simulation

Citation: Hu, B.; Xie, W.; Zhong, W.; Zhang, D.; Wang, X.; Hu, J.; Wu, Y.; Liu, Y. The Effect of Pulling Speed on the Structure and Properties of DZ22B Superalloy Blades. *Coatings* **2023**, *13*, 1225. https://doi.org/10.3390/coatings13071225

Academic Editor: Frederic Sanchette

Received: 29 May 2023
Revised: 27 June 2023
Accepted: 28 June 2023
Published: 8 July 2023

Copyright: © 2023 by the authors. Licensee MDPI, Basel, Switzerland. This article is an open access article distributed under the terms and conditions of the Creative Commons Attribution (CC BY) license (https://creativecommons.org/licenses/by/4.0/).

1. Introduction

Nickel-based superalloys have been widely used in aero engines and gas turbines to produce turbine blades with excellent high-temperature tensile strength, stress fracture, and creep properties [1–7]. DZ22B nickel-based superalloy has good casting performance, high-temperature oxidation resistance, and corrosion resistance, and is a common material for directional column crystal blades of aero engines [8–10]. However, during the solidification process, the blade is often prone to casting defects such as holes, shrinkage, and hot cracks [11,12]. Random testing of samples in different furnaces shows that the qualification rate of the samples is low, and through analysis, it is found that the main reason is defects such as slag inclusions or cracks in the samples.

Pulling speed has always been one of the most important process parameters in the directional solidification process of DZ22B superalloy blades, which affects the growth rate of dendrite tips, is closely related to the temperature gradient of the leading liquid phase of the solid-liquid interface, and also affects the occurrence of various defects in the casting process [13–20]. Therefore, the method of combining experiment and numerical simulation is used to study the effect of pulling speed on grain and dendrite growth of DZ22B superalloy blades [21], and the optimization parameters can be proposed in a targeted manner.

2. Establishment of ProCAST Numerical Simulation Model

2.1. Pre-Process Run Parameters

The casting material is DZ22B superalloy from the first author's affiliation with Aecc South Industry Company Limited, and the alloy composition is shown in Table 1. By consulting the manual [22] and combining the Procast database, the thermal physical parameters of some DZ22B superalloys are obtained as shown in Figure 1. Firstly, in order to ensure the uniformity of the composition, the DSC sample is taken from the DZ22B master alloy rather than the blade. Furthermore, a sample of about 20 mg is cut from the base metal of the DZ22B alloy, and the DSC curve is measured by using the DSC404F3 differential scanning calorimeter. As shown in Figure 1g, it can be concluded that the liquid temperature of DZ22B is 1392.6 °C and the solid temperature is 1330.9 °C. While the heat transfer coefficients of the mold shell and water-cooled crystallizer have constant values, the interfacial heat transfer coefficient between the casting and the mold shell changes with temperature according to the actual casting phase. For instance, in the solid phase range, this coefficient is 600 $W \cdot m^{-2} \cdot K^{-1}$, value that increases in the melting process to a range of 600–1500 $W \cdot m^{-2} \cdot K^{-1}$; according to the solid-liquid proportion, above the liquid phase is 1500 $W \cdot m^{-2} \cdot K^{-1}$. Other simulation parameters are shown in Table 2.

Table 1. Chemical composition of DZ22B superalloy (wt.%).

C	Cr	Co	W	Nb	Ti	Al	Hf	B	Ni
0.140	9.000	9.500	12.000	0.900	1.900	4.900	1.000	0.015	60.645

Table 2. Main simulation parameters.

Parameter (Unit)	Value
Heating zone temperature (°C)	1500
Cooling water temperature (°C)	26
Pumping speed (mm·min^{-1})	6
Heating zone emissivity	0.8
Mold shell emissivity	0.4
Mold shell density (kg·m^3)	3970
Specific heat capacity of the mold shell (KJ·kg^{-1}·K^{-1})	1.28–1.37 (1000 °C–1500 °C)
Thermal conductivity of the mold shell (W·m^{-1}·K^{-1})	9.2–5.9 (600 °C–1500 °C)
Interfacial heat transfer coefficient of Alloy and chilled plate (W·m^{-2}·K^{-1})	2400
Interfacial heat transfer coefficient of Mold shell and quench plate (W·m^{-2}·K^{-1})	1000
Interfacial heat transfer coefficient of Cooling water and chilling plate (W·m^{-2}·K^{-1})	2500

2.2. Casting and Mold Shell Model Building and Meshing

In the actual directional solidification casting production, it requires a longer time for the entire directional solidification process, which includes preheating and solidification cooling. Thus, the form of multiple blades in one mold is generally used to improve casting production efficiency.

Figure 2a is a three-dimensional model diagram of the casting of DZ22B superalloy blades with multiple sets of blades established by UG modeling software. The casting consists of a gating system, blades and a chassis, and the casting consists of 12 blades with an axisymmetric distribution as a whole.

Firstly, the model was split into quarters by using the UG modeling software because the model structure is symmetrical. Furthermore, considering the complexity of the turbine blade structure, the visual-mesh model was used to segment the surface mesh. The mesh was refined at the exhaust edge of the blade body and the small size of the truncated surface to ensure that there are enough mesh layers in the cross-sectional thickness direction, as shown in Figure 2b. Finally, the assembly of each part of the mesh was carried out after careful inspection and repair. The non-shaped shell surface and two symmetrical surfaces were specified. The thickness of the shell is set to 8mm, and the shell mesh is generated as shown in Figure 2c.

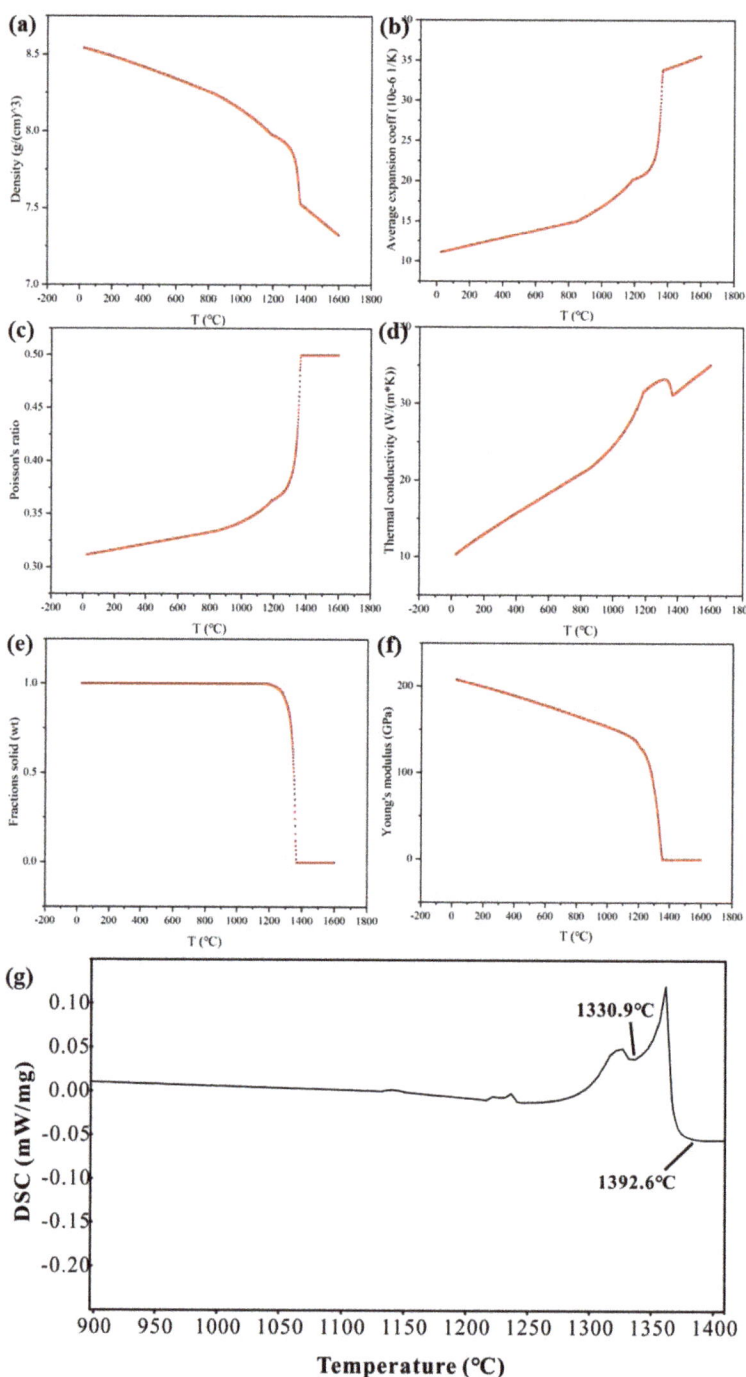

Figure 1. Thermophysical parameters as a function of temperature, (**a**) Temperature—density. (**b**) Temperature—average coefficient of expansion. (**c**) Temperature—Poisson's ratio. (**d**) Temperature—thermal conductivity. (**e**) Temperature—solid fraction. (**f**) Temperature—Young's modulus. (**g**) DSC curve.

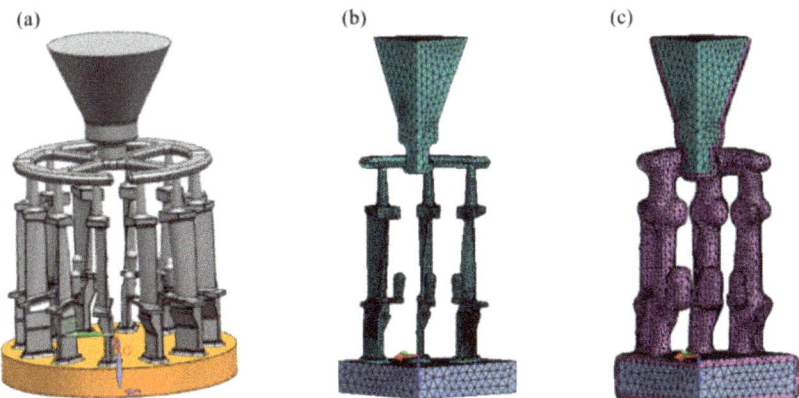

Figure 2. A multi-set blade geometry model and mesh model. (**a**) geometric model; (**b**) quarter mesh model and (**c**) shell mesh model with shell thickness of 8 mm.

2.3. Temperature Field Heat Transfer Model

In the directional solidification process, the metal pouring and solidification steps of turbine blade castings take place in a vacuum environment. In the simulation of the whole process of directional solidification, the geometry of the furnace, the shape of the blades on the quenching plate, and the pulling speed are very important. The heat transfer process of the directional solidification process can be described by the law of conservation of energy, which can be written as:

$$\rho C_p \frac{\partial(T)}{\partial(t)} = \nabla(\lambda \nabla(T)) + Q_M + Q_R \qquad (1)$$

where T is temperature; t is time; ρ is the density of matter; C_p is the specific heat of the material; λ is the thermal conductivity; Q_M is an internal heat source; Q_R is the heat flux density between the surface element and the environment. When solidifying, the liquid alloy releases latent heat, and the internal heat source can be expressed as:

$$Q_M = \rho \Delta H \frac{\partial(f_s)}{\partial(t)} \qquad (2)$$

In the formula, ΔH is the latent heat, f_s is the volume fraction of the solid phase, and the heat flux density of different directional solidification technologies is different. Using the improved Monte Carlo ray tracing model [23], the thermal radiation between the current cell and the nth reference unit is calculated as:

$$Q_R = \sigma \sum_{n=1}^{N} \frac{\beta_n (T^4 - T_n^4)}{\frac{1-\varepsilon}{\varepsilon} + \frac{(1-\varepsilon_n)S}{\varepsilon_n S_n} + 1} \qquad (3)$$

where σ is the Stefan-Boltzmann constant; "$N = 50$" is the number of rays; β_n is an energy factor; T, ε, and S are the temperature, blackness coefficient, and surface area of the surface unit, respectively. T_n, ε_n, and S_n are the surface temperature, blackness coefficient,t, and surface area of the nth element, respectively.

2.4. Grain and Dendrite Growth Models

In order to simulate grain growth during the solidification process of superalloys, an improved correspondence analysis (CA) method is used [24,25]. The total nucleation point density n(ΔT) given a supercooled ΔT is calculated using the continuous nucleation model [24]:

$$n(\Delta T) = \int_0^{\Delta T} \frac{dn}{d\Delta T'} d(\Delta T) \tag{4}$$

$$\frac{dn}{d\Delta T'} = \frac{n_{max}}{\sqrt{2\pi}\Delta T_\sigma} exp[-\frac{1}{2}(\frac{\Delta T - \Delta T_N}{\Delta T_\sigma})] \tag{5}$$

According to the Kurz-Giovanola-Trivedi (KGT) [26] kinetic equation, the growth rate of the interdendrite during solidification $v(\Delta T)$ is:

$$v(\Delta T) = \alpha \Delta T^2 + \beta \Delta T^3 \tag{6}$$

where α and β are kinetic coefficients.

2.5. Thermal Crack Susceptibility Model

In order to predict the thermal cracking sensitivity of castings, a thermal cracking sensitivity model based on Gurson's constitutive model is used [27] hot tearing indicator (HTI). The model is based on a strain-driven model of total strain during solidification. When the solids fraction is in the range of critical solids fraction (typically 50%) and 99%, the model calculates the elastoplastic strain at a given node and describes the thermal crack sensitivity of the casting by considering the initiation and growth of pores in the mushy zone of the casting. HTI is defined as follows:

$$e_{HT} = \bar{\varepsilon}_{ht}^P = \int_{t_C}^{t} \dot{\varepsilon}^P dt, t_C \leq t \leq t_S \tag{7}$$

where $\bar{\varepsilon}_{ht}^P$ is the critical cumulative effective plastic strain at the beginning of the hot cracking state, t_C is the time at the condensation temperature at which the liquid state begins to transition to a solid state, and $\dot{\varepsilon}^P$ is the effective plastic strain rate, and t_S the time at the solid phase temperature. HTI is actually a cumulative plastic strain in the semi-solid region corresponding to the pore nucleus. Therefore, it can reflect the sensitivity of hot cracking during solidification. Gurson's constitutive model and the proposed HTI model are implemented in ProCAST finite element software, and HTI should be used to explore different influencing factors of the same alloy.

3. Procast Numerically Simulates the Effect of Pulling Speed on DZ22B Superalloy Blades

3.1. Effect of Pulling Speed on Temperature Field of DZ22B Superalloy Blade

In order to investigate the effect of pulling speed on the temperature field of DZ22B superalloy blades during directional solidification while keeping other simulation parameters such as shell wall thickness, thermal conductivity of the mold shell, interfacial heat transfer coefficient, shell temperature, and insulation time unchanged. The ProCAST preprocessing module set the pulling speed to 2 mm/min, 6 mm/min, and 10 mm/min, respectively. The temperature field simulation results obtained are shown in Figure 3.

From the simulation results, the change in pulling speed will significantly affect the temperature field and mushy zone. In the early stage of pulling, the whole casting has just entered the cooling area made of the water-cooled copper ring through the thermal insulation zone; the main heat dissipation method at this time is the cooling provided by the cold plate in contact with the bottom of the casting, and the three different pulling speeds have no obvious effect on the mushy zone at the lower end of the blade. The temperature gradient is large, and the isotherm interface is very flat. As the pulling progresses, the body of the blade begins to enter the cooling zone, at which time the cooling effect provided by the cold plate is weakened due to distance, and the furnace radiation plays the main role in heat dissipation. At this time, the isotherm at the blade position showed different shapes with the change of pumping speed, and the solid-liquid interface isotherm showed an upward convex form at 2 mm/min, and the mushy zone became uneven. When the pulling speed is 6 mm/min, the solid-liquid interface isotherm turns stable, and the mushy zone tends to be horizontal. When the pulling speed is increased to 10 mm/min, the solid-liquid interface isotherm is concave, the paste area deviates significantly from the straight state, and the solidification speed of the exhaust edge position is much greater than that of the intake side. The morphology of the mushy zone affects the growth trend of grains, and the speed of 6 mm/min at the three pulling speeds is more conducive to the growth of straight columnar crystals.

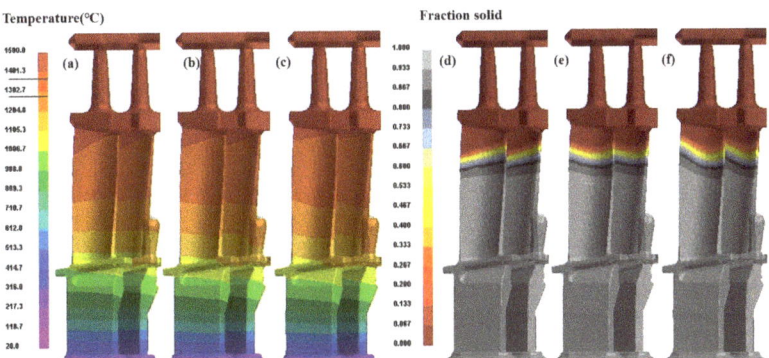

Figure 3. Schematic diagram of the temperature field and the corresponding paste zone at different pulling speeds, (**a**,**d**) 2 mm/min. (**b**,**e**) 6 mm/min. (**c**,**f**) 10 mm/min.

3.2. Effect of Pulling Speed on the Microstructure of DZ22B Superalloy Blades

Figure 4 shows the grain trend diagram at different pulling speeds, and it can be seen that the number of grains is not much different under the three pulling speeds, but the pulling speed has a certain influence on the grain direction. At low pulling speeds, the grain trend is relatively stable because the paste area is relatively stable. At the pulling speed of 10 mm/min, the paste area is concave, the grain trend has a tendency to diverge, the angle between the grain trend and the <001> direction in the microstructure is large, and the angle is close to 30°. Deviated grains can produce hetero-crystalline defects, which degrade the structure properties formed by the final solidification.

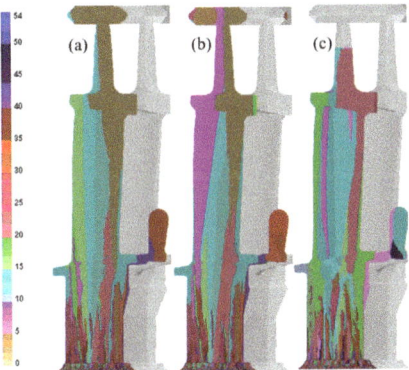

Figure 4. Schematic diagram of grain direction at different pulling speeds, (**a**) 2 mm/min. (**b**) 6 mm/min. (**c**) 10 mm/min.

The simulation results show that the dendrites have not fully grown when the pulling speed is too high. Moreover, too fast pulling speed resulted in thenges bec growth direction of the casting changes because the thermal radiation direction of the blades is uncertain and the heat loss is excessive, leading the dendrites to be tilted. Too fast pulling speed will also make the growth of columnar crystals unstable; the upper part of the blade grain will be few and divergent, and even transverse columnar crystals may appear. Only by increasing the ratio of a temperature gradient to growth rate and adjusting to determine the appropriate pumping rate can better columnar crystals be obtained.

Adjusting the pulling speed is an artificially adjustable process to effectively change the cooling rate of castings, this change will be reflected in the dendrite spacing of the solidified structure of castings, especially on the secondary dendrite arm spacing (SDAS) with the same temperature gradient. Figure 5 shows the distribution of secondary dendrite spacing at different pulling speeds, and it can be seen that the secondary dendrite spacing significantly decreases with an increase in pulling speed. At a pumping speed of 2 mm/min, the SDAS range of the blade body position is between 44–56 µm; At a pumping speed of 6 mm/min, the SDAS range of the blade body position is between 36–48 µm; When the pulling speed is increased to 10 mm/min, the SDAS drops to 32–40 µm. The secondary dendrite spacing is reduced and the microstructure performance is improved, so an appropriate increase in the pulling speed can achieve the purpose of optimizing the structure.

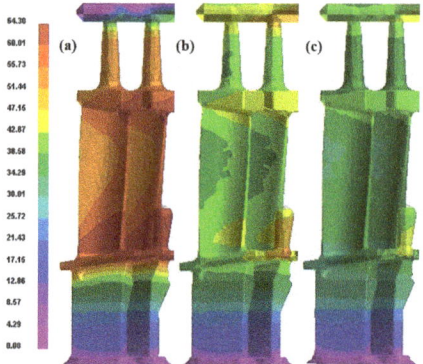

Figure 5. Secondary dendrite spacing diagram at different pull-out speeds, (**a**) 2 mm/min. (**b**) 6 mm/min. (**c**) 10 mm/min.

3.3. Effect of Pulling Speed on the Thermal Cracking Tendency of DZ22B Superalloy Blade

Figure 6 is the effective plastic strain diagram of the blade and the corresponding thermal crack index HTI diagram at different pulling speeds, from which it can be seen that changing the pulling speed has a certain effect on the plastic strain and thermal cracking sensitivity of the blade after solidification. The chance of thermal cracking of the blade is lowest at a pulling speed of 6 mm/min, and only a small number of hot crack-sensitive areas are located at the junction of the upper part of the blade body with the blade crown. However, the thermal cracking sensitivity of the blade formed by directional solidification at lower or higher pulling speeds is large, and hot cracks may be formed in many places. The effective plastic strain value obtained by simulation has a certain relationship with the pulling speed, and the effective plastic strain of the blade at the pulling speeds of 2 mm/min and 10 mm/min is significantly greater than the pulling speed of 6 mm/min.

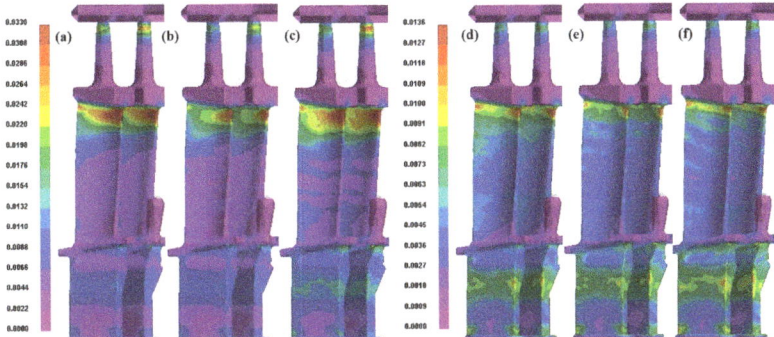

Figure 6. Effective plastic strain diagram of the blade at different pulling speeds and HTI diagram of the thermal crack index, (**a**,**d**) 2 mm/min. (**b**,**e**) 6 mm/min. (**c**,**f**) 10 mm/min.

3.4. Experimental Verification

Experiments are conducted on blades with hot cracks produced by AECC Southern Industry Company Limited. The cracks are detected using fluorescence detection, while the macroscopic grain distribution is observed after the sample is corroded with hydrochloric acid, and it is found that the hot crack is generated in the middle and upper part of the exhaust side of the blade body; usually a crack penetrates the blade body. The fluorescence detection and macroscopic grain corrosion results are shown in Figure 7. The pulling speeds are 2 mm/min, 6 mm/min, and 10 mm/min for DZ22B superalloy blades, which are the same as those actually produced, and cross-sectional samples are prepared by taking the middle and upper positions of the blade body. SEM images of the blade pores after the corrosion process are obtained at various pulling speeds (Figures 8 and 9).

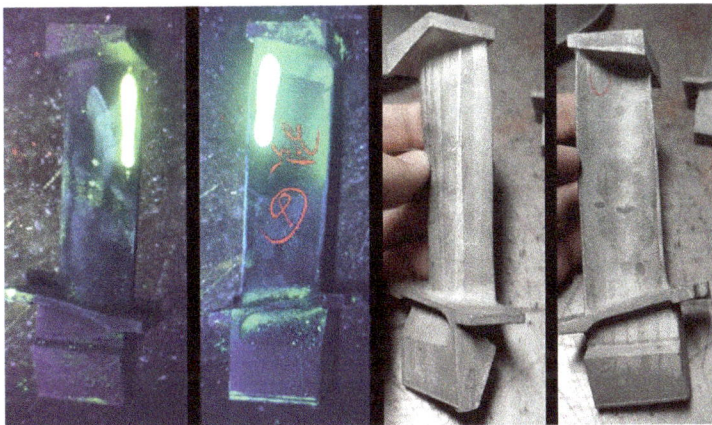

Figure 7. The fluorescence detection and macroscopic grain corrosion of the hot cracked blade, and the red circle is the hot crack.

By observing the sample SEM diagram at different pulling speeds, it can be seen that as the pulling speed increases, the secondary dendrite spacing is significantly shortened, and the secondary dendrite spacing is the largest at the pulling speed of 2 mm/min, so its cracking tendency will be higher. The dendrite spacing depends mainly on the temperature gradient G_L and the solidification rate R, the product of which is called the cooling rate, so ($G_L \cdot R$) determines the size of the dendrite spacing as follows:

$$\lambda_1 = [-64\theta D_L m_L (1-k) C_0]^{1/4} G_L^{-1/2} R^{-1/4} \tag{8}$$

$$\lambda_2 = B(G_L V)^{-1/3} \tag{9}$$

Figure 8. Cylindrical specimen mid-pore SEM diagram at different pull-out speeds, (**a**) 2 mm/min. (**b**) 6 mm/min. (**c**) 10 mm/min.

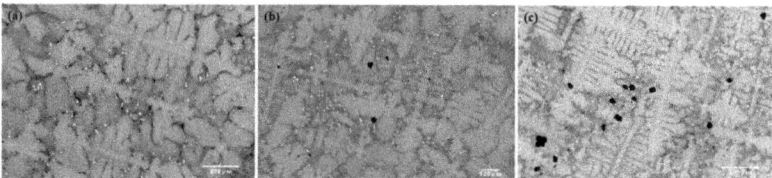

Figure 9. SEM diagram of the upper porosity of a specimen at different pulling speeds, (**a**) 2 mm/min. (**b**) 6 mm/min. (**c**) 10 mm/min.

In which, λ_1 is the primary dendrite spacing; λ_2 is the secondary dendrite spacing; θ is the capillary constant; G_L is the temperature gradient at the front edge of the solid-liquid interface; k is the solute partition coefficient; C_0 is the concentration of alloying elements; D_L is the diffusion coefficient of solute elements in the liquid phase; m_L is the slope of the liquid line; k is the solidification rate; B is a constant; V is the pulling speed. It can be seen that for superalloys with fixed composition, except for G_L and R, the other quantities have fixed values. Primary dendrite spacing, secondary dendrite spacing, and cooling rate ($G_L \cdot R$) satisfy a positively correlated relationship. In general, the increase in the pulling speed makes the temperature gradient of the liquid phase smaller but not large, the overall cooling rate ($G_L \cdot R$) increases, and the distance between primary and secondary dendrites decreases. In addition, with the continuous improvement of the pulling speed, the cooling rate of the blades accelerates, and the heat dissipation effect becomes better and better, resulting in the refinement of the cast dendrite structure, so the crystalline size also decreases. Therefore, the tendency for thermal cracking can be reduced to a certain extent by increasing the pulling speed.

The number of shrinkage holes at the three pulling speeds in the middle of the sample was relatively small; fine shrinkage holes were distributed between the dendrites at the pulling speeds of 2 and 6 mm/min, and irregular shrinkage holes were clustered together at the 10 mm/min pulling speed, and the pore length reached 160.2 μm. By observing the pore SEM diagram of three different pumping speeds in the upper part of the specimen, it can be found that with the increase in pulling speed, the number of pores increases significantly, and the shrinkage holes have aggregation at the pulling speed of 10 mm/min. The percentage of porosity at 2 mm/min, 6 mm/min, and 10 mm/min pulling speeds is calculated by Image-pro, which is 0.079%, 0.239%, and 0.643%, respectively; the faster the pulling speed, the more pores in the upper part of the test bar. When the pulling speed is 10 mm/min, although the transverse dendrite group obtained is small, it can be seen that the dendrite orientation is disordered and the occlusion of the longitudinal secondary dendrite is inadequate. The reason for this phenomenon is that the growth rate of the crystal is lower than the pulling rate, the solid-liquid interface growth front is below the insulation plate, the solidification environment of the axial one-dimensional heat flow changes, and the heat loss occurs along the transverse direction, resulting in dendrite orientation disorder. At the same time, a large pulling speed will increase the solidification rate of the blades and the volume of solidified solids per unit of time, thus releasing more heat from the latent heat of crystallization. Furthermore, it will decrease the negative temperature gradient parallel to the growth direction of the secondary dendrite and reduce the growth driving force, resulting in insufficient occlusion of the secondary dendrite. In addition, when the pulling speed is too fast, the radiation heat dissipation will deviate from the directional solidification direction, which can easily lead to oblique crystals and crystal-breaking defects. In short, excessive pulling speed will increase the tendency of alloy hot cracking.

In summary, with the increase in pulling speed, the cooling rate during the directional solidification of castings gradually becomes faster, while the primary and secondary dendrite spacing of the alloy decreases. When the pulling speed of directional solidification is too small, although the temperature gradient is improved, the cooling rate of the casting is slow, the dendrite spacing is large, and the solidification segregation is significant. When the directional solidification pulling speed is too large, although the cooling speed is fast and the dendrite spacing is small, it is not conducive to the sequential solidification of the casting, the secondary dendrite bite is insufficient, and the tendency to form micro constriction is increased; therefore, in order to reduce the enthusiastic tendency of directional solidification, a moderate pulling rate should be selected, and the experimental results are in good agreement with the simulation results.

4. ProCAST Numerically Simulates the Effect of Variable Speed Pulling Process on DZ22B Superalloy Blades

Turbine blades are often prepared at a single pulling rate in the industrial production of directional solidification. In the previous work, the influence of a constant pulling rate on the shape and temperature gradient of the solid-liquid interface, as well as the influence on the structural performance of blades, were systematically analyzed. Appropriate pulling speed can reduce the secondary dendrite spacing of microstructure, improve casting performance, reduce pore formation, and reduce the thermal crack sensitivity index. The uniform speed pulling law obtained above is applied to the exploration of the variable speed pulling process: the initial pulling rate is 6 mm/min, the pulling height is 110 mm, the later variable speed is 2 mm/min or 10 mm/min, and the pulling height is not less than 180 mm, and the pulling speed scheme of different parts of the blade is obtained as shown in Figure 10.

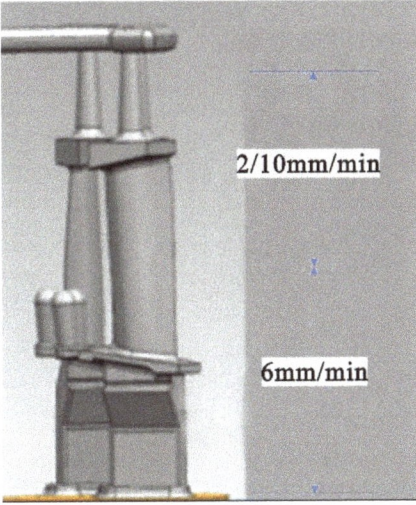

Figure 10. Variable speed pulling scheme.

After measuring the blade length in advance, taking the position of the heat insulation baffle as the reference point, the pulling speed of 6 mm/min is adopted in the early stage because the alloy liquid is affected by the excited chill plate during the directional solidification in the early stage, the cooling gradient is very large, and the mushy area moves faster. The variable speed is realized below the marginal plate, and the effect of variable speed pulling on the hot cracking tendency of DZ22B alloy is studied from 6 mm/min to 2 mm/min or 10 mm/min.

4.1. Effect of Variable Speed Pulling on Temperature Field of DZ22B Superalloy Blade

The pulling speed has a great influence on the temperature gradient of the liquid phase at the front edge of the solid-liquid interface, and the regulation of the variable speed pulling process is directly reflected in the evolution of the temperature field and the morphological change of the mushy zone during the directional solidification process of the casting. The temperature field comparison chart of the 6–2 mm/min and 6–10 mm/min variable speed pulling processes is shown in Figures 11 and 12; it is the comparison chart of the solidified mushy area of 6–2 mm/min and 6–10 mm/min variable speed pulling processes.

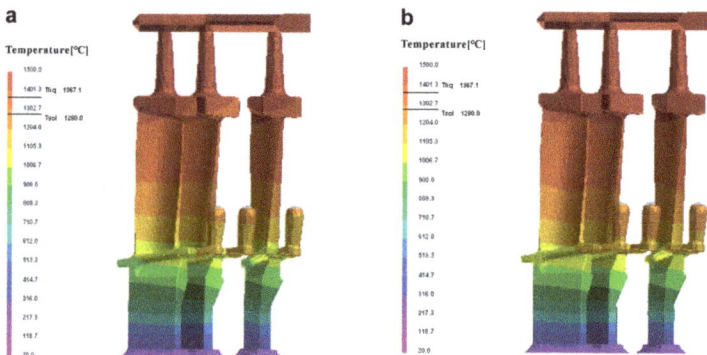

Figure 11. ProCAST simulation of variable speed pulling directional solidification temperature field results, (**a**) 6–2 mm/min. (**b**) 6–10 mm/min.

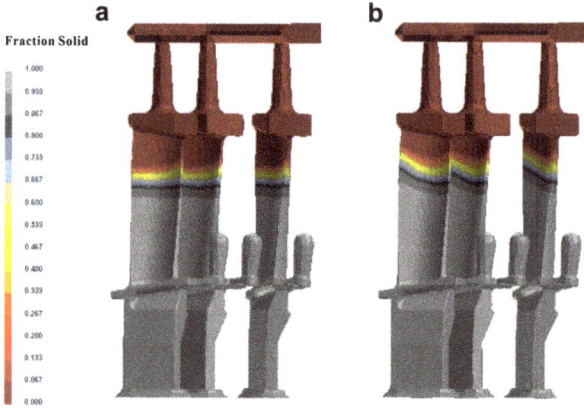

Figure 12. ProCAST simulated variable speed pulling directional solidification mushy zone morphology, (**a**) 6–2 mm/min. (**b**) 6–10 mm/min.

When the mushy area is in the upper position, the casting has just completed the deceleration pulling of 6 mm/min–2 mm/min or the accelerated pulling of 6 mm/min–10 mm/min, it can be seen that the mushy area of the upper part of the blade in Figure 12a is relatively flat, while the mushy area of the upper part of the blade in Figure 12b is concave, which will affect the solidification behavior of the upper part of the blade, making the solidification structure uneven and easy to form uneven stress, resulting in hot cracks.

4.2. Effect of Variable Speed Pulling on the Thermal Cracking Tendency of DZ22B Superalloy Blade

Figures 13 and 14 show the effective plastic strain diagram of the variable speed pulling blade and the corresponding thermal crack index HTI diagram, in which it can be seen that the plastic strain is concentrated in the upper position of the blade. It can be seen that the strain diagram of 6 mm/min slowing down to 2 mm/min or accelerating to 10 mm/min is obtained by decelerating the pulling speed to 2 mm/min or accelerating to 10 mm/min at the upper part of the blade, and the stress gap in the upper part of the blade is particularly obvious compared with the strain diagram accelerated to 10 mm/min. When the pulling rate accelerates to 10 mm/min, the cooling is accelerated due to the increase in speed, which increases the internal stress and finally leads to the generation of thermal cracking.

The color bars and corresponding numbers on the left side of the simulated image indicate the trend of the simulation parameters. In Figure 14, purple represents a small tendency for thermal cracking, while red represents a large tendency for thermal cracking. It can be seen from Figure 14 that the variable speed pulling rate of 6 mm/min–10 mm/min is larger than the variable speed pulling rate of 6 mm/min–2 mm/min, and the dark blue region is larger and the purple region is smaller, indicating that the area with a large thermal cracking tendency is wider. According to the simulation results of uniform pulling, it can also be seen that when the pulling rate is constant, the hot cracking tendency of the upper edge of the last solidified blade is greater, and the pulling rate should be reduced so that the last solidified part can be fully compensated, thereby reducing the thermal cracking tendency. Therefore, from the perspective of reducing the tendency of thermal cracking, a variable speed pulling rate of 6 mm/min–2 mm/min should be selected.

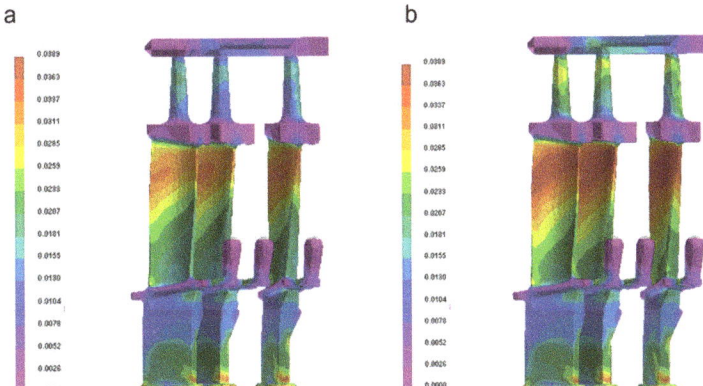

Figure 13. Comparison of the equivalent force of variable speed pulling blades, (**a**) 6–2 mm/min. (**b**) 6–10 mm/min.

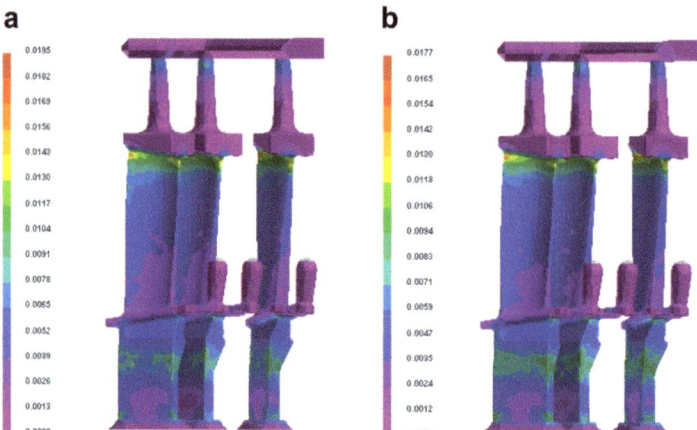

Figure 14. Comparison of the thermal cracking tendency of variable speed pulling blades, (**a**) 6–2 mm/min. (**b**) 6–10 mm/min.

4.3. Experimental Verification

In order to study the effect of variable speed pulling rate on DZ22B directional solidification single crystal blades, three experiments with different variable speed pulling rates are designed, namely: (a) the first 110 mm crystal pulling speed is 6 mm/min, then reduced to 2 mm/min, the pulling height is not less than 180 mm. (b) The front and rear pulling rates remain unchanged, and the pulling speed is 6 mm/min. (c) The first 110 mm crystal pulling speed is 6 mm/min and then rises to 10 mm/min, and the pulling height is not less than 180 mm.

Through fluorescence non-destructive testing, no cracks were found at these three pulling rates, indicating that the change in pulling rate would not result in cracks. As shown in Figure 15, microcracks can be observed in the microstructures whose pulling rates are different. However, it should be noted that the crack size is significantly larger when the pulling rate is constant at 6 mm/min. Therefore, variable speed pulling can effectively reduce microcracks, thereby reducing the tendency for hot cracking.

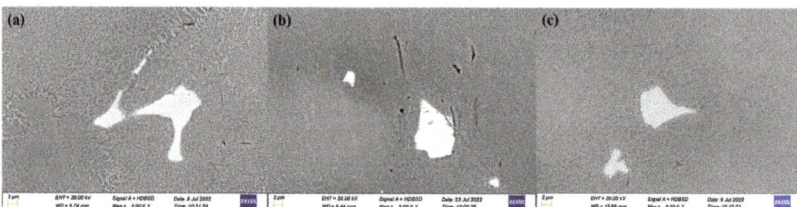

Figure 15. SEM diagram of precipitated phase in the middle of the air intake side, (**a**) 6–2 mm/min. (**b**) 6–6 mm/min. (**c**) 6–10 mm/min.

Figure 16 shows the sampling position of the secondary dendrite spacing, no obvious cracks were found on the cut surface of each group of blades after sampling, and Figure 17 shows the secondary dendrite photos of the upper position of the blades with different pulling rates, and there is no obvious change in the secondary dendrite spacing in the area where the speed has changed. However, it can be observed that when the pulling rate is 6–10 mm/min, the dendrite orientation is disordered and the longitudinal secondary dendrite bite is insufficient. The microstructure of different pulling rates is shown in Figure 18: it can be found that the structure size is more uniform at the pulling rate of

6–2 mm/min, there is no abnormally developed secondary dendrite, and the grain trend is relatively stable.

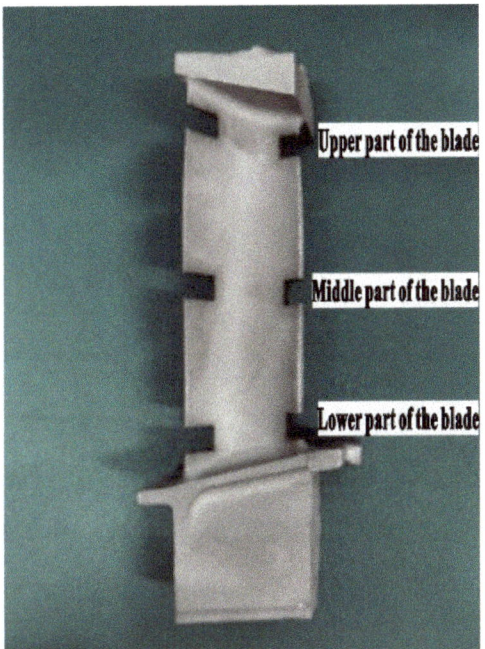

Figure 16. Sampling position of secondary dendrite spacing.

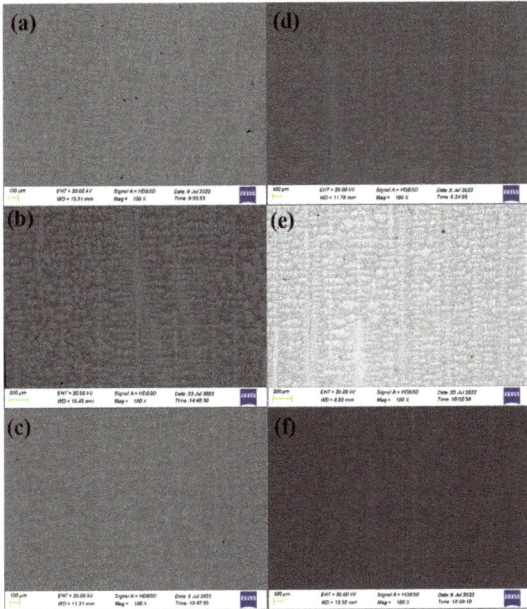

Figure 17. Secondary dendrite SEM diagram at the upper position of the intake end, (**a**) 6–2 mm/min. (**b**) 6–6 mm/min. (**c**) 6–10 mm/min. Secondary dendrite SEM diagram at the upper position of the exhaust end, (**d**) 6–2 mm/min. (**e**) 6–6 mm/min. (**f**) 6–10 mm/min.

When the pulling speed is too high, the shape of the mushy area is concave, which is not conducive to the formation of uniform, straight columnar crystals. Under the three pulling speeds, the grain orientation with lower and medium pulling speeds is better. The structure size is more uniform at the pulling speed of 6–2 mm/min, and there is no abnormally developed secondary dendrite; while the grain trend is relatively stable, the higher the pulling speed, the greater the chance of forming pores when the casting solidifies in the later stage, and the formation of pores may be the cause of thermal cracking. Therefore, a pulling rate of 6–2 mm/min should be selected when changing the speed of pulling.

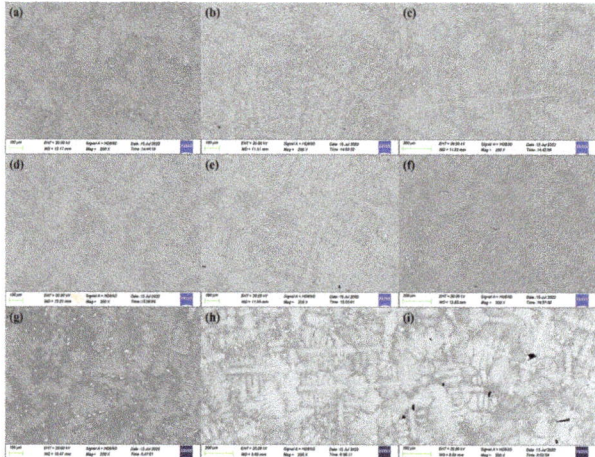

Figure 18. 6–2 mm/min pulling speed, blade cross-section, SEM diagram, (**a**) lower part of the blade. (**b**) the middle part of the blade. (**c**) the upper part of the blade. 6–6 mm/min pulling speed, blade cross-section, SEM diagram, (**d**) lower part of the blade. (**e**) the middle part of the blade. (**f**) the upper part of the blade. 6–10 mm/min pulling speed, blade cross-section, SEM diagram, (**g**) lower part of the blade. (**h**) the middle part of the blade. (**i**) the upper part of the blade.

5. Conclusions

DZ22B alloy is commonly used as a blade material for aircraft engines and gas turbines, and its preparation process is an important factor affecting its performance. In the present work, a reliable numerical model has been established through ProCast numerical simulation that assumes a set of fixed operating and thermo physical parameters and complemented with an auxiliary experimental verification method. The effect of casting speed on the grain and dendrite growth of DZ22B superalloy blades is studied. Based on numerical simulation and experimental verification, the following conclusions can be drawn:

(a) The simulation results show that the solid-liquid interface isotherm is convex at a pulling speed of 2 mm/min, the solid-liquid interface is concave at a pulling speed of 10 mm/min, the solid-liquid interface isotherm changes smoothly at a pulling speed of 6 mm/min.

(b) The morphology of the mushy zone affects the growth trend of grains, and the speed of 6 mm/min is more conducive to the growth of straight columnar crystals, which is consistent with the experimental results.

(c) Variable-speed pulling can effectively reduce micro-cracks, thereby reducing the tendency for hot cracking. At the pulling speed of 6–2 mm/min, the grain size is more uniform, and there are no abnormally developed secondary dendrites.

(d) Based on the above results, it is recommended to select 6–2 mm/min variable pulling speed for the directional solidification of the DZ22B superalloy blade.

Author Contributions: Conceptualization, B.H.; Methodology, W.X.; Validation, W.Z. and X.W.; Writing—original draft, D.Z.; Writing—review & editing, J.H.; Supervision, Y.W. and Y.L. All authors have read and agreed to the published version of the manuscript.

Funding: This work was financially supported by the Provincial Department of Education Project (Nos. 21C0090).

Institutional Review Board Statement: Not applicable.

Informed Consent Statement: Not applicable.

Data Availability Statement: Not applicable.

Conflicts of Interest: The authors declare no conflict of interest.

References

1. Xiao, X. Thermal fatigue behaviors of three cast nickel base superalloys. *Acta Met. Sin* **2011**, *47*, 1129–1134.
2. Khan, M.A. Oxidation and hot corrosion behavior of nickel-based superalloy for gas turbine applications. *Mater. Manuf. Process* **2014**, *29*, 832–839. [CrossRef]
3. Zhang, S. Anisotropie creep rupture properties of a nickel-base single crystal superalloy at high temperature. *J. Mater. Sci. Technol.* **2011**, *27*, 107–112. [CrossRef]
4. Liu, C. Effect of withdrawal rates on microstructure and creep strength of a single crystal superalloy processed by LMC. *J. Mater. Sci. Technol.* **2010**, *26*, 306–310. [CrossRef]
5. Han, G. Effect of threshold stress on anisotropic creep properties of single crystal nickel-base superalloy SRR99. *J. Mater. Sci. Technol.* **2012**, *28*, 439–445. [CrossRef]
6. Liu, Y. Interaction between CA6-MA crucible and molten wrought Ni-based superalloys. *J. Eur. Ceram. Soc.* **2023**, *43*, 1714–1722. [CrossRef]
7. Su, S.A. The formation and evolution of stray grains in remelted interface in the seed crystal during the directional solidification of single crystal superalloys assisted by vertical static magnetic field. *J. Acta Met. Sin* **2022**. [CrossRef]
8. Li, F. Investigation of fused alumina based-mold facecoats for DZ22B directionally solidified blades. *Materials* **2019**, *12*, 606. [CrossRef]
9. Liu, Y.N. Phase Transformation Process and Solidification Microstructure Transition of DZ22B superalloy. *J. East China Jiaotong Univ.* **2018**, *35*, 99–104.
10. Li, Z. Effect of Slurry Thickness on the Quality of Aluminized Coatings. *J. Mater.* **2022**, *15*, 6758. [CrossRef]
11. Yan, X. Investigation on material's fatigue property variation among different regions of directional solidification turbine blades—Part I: Fatigue tests on full scale blades. *J. Eng. Gas Turbine Power* **2014**, *136*, 102502. [CrossRef]
12. Xu, Z. Experimental study on mechanical properties of silica-based ceramic core for directional solidification of single crystal superalloy. *Ceram. Int.* **2018**, *44*, 394–401. [CrossRef]
13. Kanyo, J.E. An overview of ceramic molds for investment casting of nickel superalloys. *J. Eur. Ceram. Soc.* **2020**, *40*, 4955–4973. [CrossRef]
14. Du, B. Investigation on the microstructure and tensile behavior of a Ni-based IN792 superalloy. *Adv. Mech. Eng.* **2018**, *10*, 1687814017752167. [CrossRef]
15. Zhao, Y. Phase-field simulation for the evolution of solid/liquid interface front in directional solidification process. *J. Mater. Sci. Technol.* **2019**, *35*, 1044–1052. [CrossRef]
16. Shen, Y. Micro-segregation study of directional solidified Fe-Mn-C-Al TWIP steels. *Met. Mater. Trans.* **2020**, *51*, 2963–2975. [CrossRef]
17. Zhu, C.S. Multi-phase field simulation of competitive grain growth for directional solidification. *Chin. Phys. B* **2022**, *31*, 068102. [CrossRef]
18. Zhu, K. Mechanism of primary Si and Fe-segregation for a hypereutectic 90 wt.% Si-Ti alloy melt in directional solidification via electromagnetism. *J. Iron Steel Res. Int.* **2023**, *30*, 726–735. [CrossRef]
19. Gao, Z. Multi-phase field model simulation based on MPI+ OpenMP parallel: Evolution of seaweed and dendritic structure in directional solidification. *J. AIP Adv.* **2022**, *12*, 035018. [CrossRef]
20. Fang, H. Effects of Heating Power on Microstructure Evolution and Tensile Properties at Elevated Temperature by Directional Solidification for Ti2AlC/TiAl Composites. *J. Adv. Eng. Mater.* **2022**, *24*, 2100736. [CrossRef]
21. Yu, F. the influence of anisotropy on the evolution of interfacial morphologies in directional solidification: A phase-field study. *arXiv* **2022**, arXiv:2210.15673.
22. The editorial board of china aeronautical materials handbook. DZ22B; In *China Aeronautical Materials Handbook*; Publishing House: Beijing, China, 2002; pp. 751–757.
23. Qingyan, X. Progress on modeling and simulation of directional solidification of superalloy turbine blade casting. *China Foundry* **2012**, *9*, 69–77.

24. Rappaz, M. Probabilistic modelling of microstructure formation in solidification processes. *Acta Metall.* **1993**, *41*, 345–360. [CrossRef]
25. Gandin, C. A three-dimensional cellular automation-finite element model for the prediction of solidification grain structures. *Met. Mater. Trans.* **1999**, *30*, 3153–3165. [CrossRef]
26. Kurz, W. Theory of microstructural development during rapid solidification. *Acta Metall.* **1986**, *34*, 823–830. [CrossRef]
27. Zhu, J. Numerical modeling of hot tearing formation in metal casting and its validations. *Int. J. Numer. Methods Eng.* **2011**, *87*, 289–308. [CrossRef]

Disclaimer/Publisher's Note: The statements, opinions and data contained in all publications are solely those of the individual author(s) and contributor(s) and not of MDPI and/or the editor(s). MDPI and/or the editor(s) disclaim responsibility for any injury to people or property resulting from any ideas, methods, instructions or products referred to in the content.

Article

Mechanical and Tribological Behaviors of Hot-Pressed SiC/SiC$_w$-Y$_2$O$_3$ Ceramics with Different Y$_2$O$_3$ Contents

Shaohua Zhang [1,2,3], Jinfang Wang [2,3], Meng Zhang [2,3], Longqi Ding [2,3], Huijun Chan [2,3], Xiyu Liu [2,3], Fengqing Wu [2,3], Zhibiao Tu [2,3], Ling Shao [2,3], Nengyong Ye [2,3], Sheng Dai [2,3], Liu Zhu [2,3,*] and Shichang Chen [1,*]

1. School of Materials Science and Engineering, Zhejiang Sci-Tech University, Hangzhou 310018, China
2. School of Materials Science and Engineering, Taizhou University, Taizhou 318000, China
3. Zhejiang Provincial Key Laboratory for Cutting Tools, Taizhou University, Taizhou 318000, China
* Correspondence: zhuliu@tzc.edu.cn (L.Z.); scchen@zstu.edu.cn (S.C.)

Citation: Zhang, S.; Wang, J.; Zhang, M.; Ding, L.; Chan, H.; Liu, X.; Wu, F.; Tu, Z.; Shao, L.; Ye, N.; et al. Mechanical and Tribological Behaviors of Hot-Pressed SiC/SiC$_w$-Y$_2$O$_3$ Ceramics with Different Y$_2$O$_3$ Contents. *Coatings* 2023, 13, 956. https://doi.org/10.3390/coatings13050956

Academic Editor: Csaba Balázsi

Received: 18 April 2023
Revised: 8 May 2023
Accepted: 15 May 2023
Published: 19 May 2023

Copyright: © 2023 by the authors. Licensee MDPI, Basel, Switzerland. This article is an open access article distributed under the terms and conditions of the Creative Commons Attribution (CC BY) license (https://creativecommons.org/licenses/by/4.0/).

Abstract: Sintering additives are commonly used to reduce the conditions required for densification in composite ceramics without compromising their performances simultaneously. Herein, SiC/SiC$_w$-Y$_2$O$_3$ composite ceramics with 10 vol.% SiC whiskers (SiC$_w$) and different Y$_2$O$_3$ contents (0, 2.5, 5, 7.5, and 10 vol.%) were fabricated by hot-pressed sintering at 1800 °C, and the effects of Y$_2$O$_3$ content on the microstructure, mechanical properties, and tribological behaviors were investigated. It was found that the increased Y$_2$O$_3$ content can promote the densification of SiC/SiC$_w$-Y$_2$O$_3$ composite ceramics, as evidenced by compact microstructure and increased relative density. The Vickers hardness, fracture toughness, and flexural strength also increased when Y$_2$O$_3$ content increased from 2.5 vol.% to 7.5 vol.%. However, excessive Y$_2$O$_3$ (10 vol.%) aggregated around SiC and SiC$_w$ weakens its positive effect. Furthermore, the Y$_2$O$_3$ additive also reduces the coefficient of friction (COF) of SiC/SiC$_w$-Y$_2$O$_3$ composite ceramics, the higher the Y$_2$O$_3$ content, the lower the COF. The wear resistance of SiC/SiC$_w$-Y$_2$O$_3$ composite ceramics is strongly affected by their microstructure and mechanical properties, and as-sintered SiC ceramic with 7.5 vol.% Y$_2$O$_3$ (Y075) shows the optimal wear resistance. The relative density, Vickers hardness, fracture toughness, and flexural strength of Y075 are 97.0%, 21.6 GPa, 7.7 MPa·m$^{1/2}$, and 573.2 MPa, respectively, the specific wear rate of Y075 is 11.8% of that for its competitor with 2.5 vol.% Y$_2$O$_3$.

Keywords: SiC/SiC$_w$-Y$_2$O$_3$ ceramics; hot-pressed sintering; Y$_2$O$_3$ additive; mechanical properties; tribological behaviors

1. Introduction

Owing to their excellent mechanical properties and abrasion resistance, silicon carbide (SiC) ceramics have been widely used in aerospace, petrochemical, and other fields [1,2]. However, the strong covalent bonding and low self-diffusion coefficient of SiC ceramics make it hard to sinter completely and get compacted microstructure [3]. Commonly, pure SiC ceramics require at least 2000 °C to get thoroughly compacted, and this considerable energy consumption should be taken seriously [1]. Besides, the high hardness of SiC also means its high brittleness, which may not be adaptable in some complex environments. Thus, it is necessary to improve the comprehensive performance of SiC ceramic against harsh work conditions.

Whisker strengthening is a very effective method to reduce the brittleness of hard ceramics. The commonly used reinforcements include SiC fibers [4], Si$_3$N$_4$ whiskers [5], SiC whiskers (SiC$_w$) [6–9], and others. SiC$_w$ has better compatibility and has been widely used in the toughening study of SiC ceramics, owing to the same chemical composition and similar crystalline structure between SiC and SiC$_w$. SiC$_w$ reinforcement has been used to enhance the mechanical properties, particularly the fracture toughness of SiC ceramics, due to its high elastic modulus and strength. The main toughening mechanisms include whisker bridging, pullout, crack deflection, and their combination [6–9]. Several articles [6,7] mentioned that with SiC_w content increase, the mechanical properties of SiC ceramics increase first and then decrease. Yang et al. [8] reported that the proper content of SiC_w reinforcements could enhance the wear resistance, which was ascribed to the improved mechanical properties of SiC ceramics. However, few studies on SiC_w-reinforced SiC ceramics fabricated by hot-pressed sintering have been reported.

Recently, several articles have shown that SiC powders can be densified by incorporating sintering additives at a lower sintering temperature without compromising their performance [10–17]. Therefore, various additives like Y_2O_3 [10], La_2O_3 [11], and $RE(NO_3)_3$ [12] (RE = rare earth element), etc., as well as the combination of Al_2O_3-RE_2O_3 [13–15], AlN-RE_2O_3 [16,17], and Er_2O_3-Y_2O_3 [18] have been used in the sintering of SiC ceramics to improve its sinterability by liquid phase generation. Further, the Y_2O_3 additive has been used in various ceramics, including SiC [19–22], Si_3N_4 [23,24], TiC [25], ZrB_2-SiC [26–30], AlN-SiC [31], WC [32–34], and cordierite [35], etc. Some references [19–35] mentioned that the Y_2O_3 additive could promote the densification of the materials and influence their mechanical properties. Cheng et al. [25] presented that Y_2O_3 can improve the mechanical properties of TiC ceramics by pinning at the grain boundary. Zhang et al. [26] discovered that appropriate Y_2O_3 addition could eliminate surface oxides and suppress grain growth to promote the densification of ZrB_2-SiC ceramics. Several articles [10,19–22] reported that the Y_2O_3 additive could improve the densification of SiC ceramics at a higher sintering temperature (1900–2000 °C) to achieve excellent performance. It demonstrates that Y_2O_3 additions can accelerate the diffusion and mass transfer between the SiC matrix during sintering [12,20]. However, a much higher temperature can deteriorate the mechanical properties by vaporizing the liquid phase and generating pores [36]. Gupta et al. [37] reported that a small amount of Y_2O_3 addition could improve the mechanical and tribological properties of β-SiC ceramics. However, the preparation of SiC/SiC_w-Y_2O_3 composite ceramics and the effect of Y_2O_3 content on the mechanical and tribological properties have rarely been reported.

To obtain compact SiC ceramics at a comparatively lower sintering temperature and improve their mechanical and tribological properties simultaneously, Y_2O_3 and SiC_w are incorporated to fabricate the SiC/SiC_w-Y_2O_3 composite ceramics by hot-pressed sintering (HP) at 1800 °C. SiC_w is used as the reinforcement and kept at 10 vol.%. Y_2O_3 is the sintering additive, with content increasing from 0 to 10 vol.%. The effects of Y_2O_3 contents on the microstructure, mechanical properties, and tribological behaviors of the SiC/SiC_w-Y_2O_3 composite ceramics were investigated. This study can provide constructive suggestions and references to the field of SiC ceramic modification.

2. Experimental

2.1. Raw Powders

Alpha-SiC nanoparticles (≥99.9%, ~100 nm, Forsman, Beijing, China), β-SiC_w (≥99.9%, D0.5 μm-L12 μm, Forsman, Beijing, China), and Y_2O_3 nanoparticles (≥99.9%, ~50 nm, Aladdin, Shanghai, China) were used in this study to sinter the SiC/SiC_w-Y_2O_3 ceramics. FE-SEM (scanning electron microscopy, S-4800, HITACHI, Tokyo, Japan) images of the 3 raw powders and their XRD (X-ray diffractometer, D8 Advance, Bruker, Billerica, MA, USA) patterns are shown in Figure 1. SiC powders are irregular polygonal particles with a main crystalline structure of 6H-SiC (PDF-# 72-0018). SiC_w has a diameter of ~0.5 μm and a main crystalline structure of 3C-SiC (PDF-# 75-0254). Y_2O_3 (PDF-# 65-3178) is a subspindle floccule and agglomerated with adjacent ones. The obtained results confirmed the satisfactory morphology, size, and crystallinity of the raw powders.

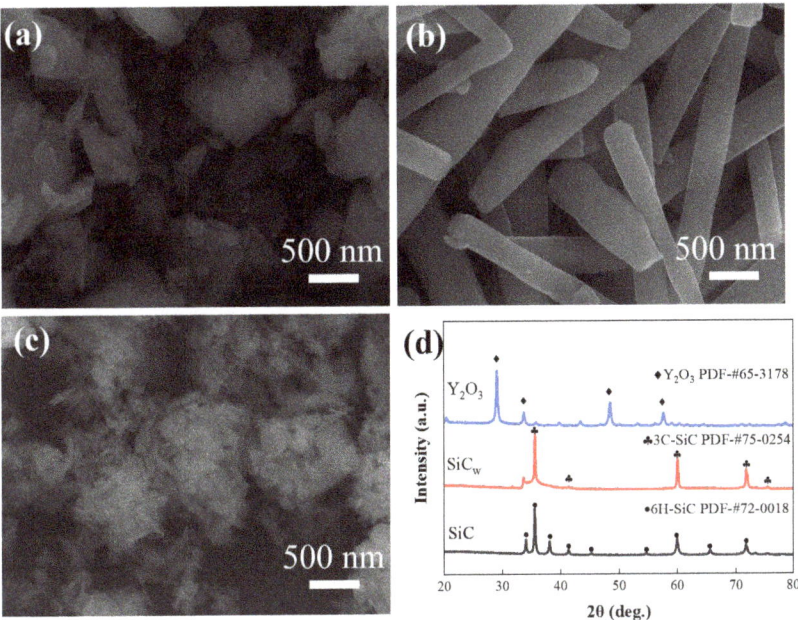

Figure 1. SEM images of (**a**) SiC powder, (**b**) SiC$_w$ powder, (**c**) Y$_2$O$_3$ powder, and (**d**) the XRD patterns of the raw powders.

2.2. Hot-Pressed Sintering of SiC/SiC$_w$-Y$_2$O$_3$ Ceramics

To explore the effect of Y$_2$O$_3$ content on the microstructure and comprehensive properties of SiC/SiC$_w$-Y$_2$O$_3$ ceramics, Y$_2$O$_3$ with volume content of 0%, 2.5%, 5%, 7.5%, and 10% (denoted as Y000, Y025, Y050, Y075, and Y100, respectively) were investigated in this study. SiC$_w$ contents were kept constant (10 vol.%), and SiC contents were in balance. The purchased powders were mixed according to their designed formula, then with ZrO$_2$ balls of 10 times the weight of the obtained powder and ethanol in the polyamides jar. The ball-milling of the raw powders was conducted by a planet-ball-grinding machine (MIRT-QMQX-4L, Miqi, Changsha, China), and the mill lasted without intervals for 12 h with a speed of 180 r/min. The ball-milled powders were dried, sieved, and pre-compacted in a graphite mold coated with graphite paper. Then, the green compacts were fabricated by a vacuum hot-pressed sintering furnace (ZT-40-21Y, Chenhua, Shanghai, China) to obtain SiC/SiC$_w$-Y$_2$O$_3$ composite ceramics with dimensions of φ26.5 mm × 6.5 mm. The sintering process was conducted at 1800 °C for 1 h under the pressure of 40 MPa in a vacuum atmosphere. The detailed sintering procedure can be found in Figure 2.

2.3. Sample Characterization

The as-sintered ceramics were polished by the diamond polishing disc and silk flannelette. The diamond spray with sizes gradually decreased to 0.25 μm was used during polishing. The specimen surfaces were finally polished to average roughness below 0.4 μm. The phase compositions of the as-sintered ceramics were detected by the mentioned XRD. Before XRD detection, the composite ceramics were carefully polished. The fracture morphologies of the composite ceramics were obtained by the three-point bending method and observed by the SEM (S-4800, HITACHI). The morphologies and chemical composition of the polished ceramics were examined by a photo-diode back-scattered electron (PDBSE) detector and an energy-dispersive X-ray spectrum (EDS, X-Max^N 20, Oxford, Oxford, UK) detector equipped on the SEM.

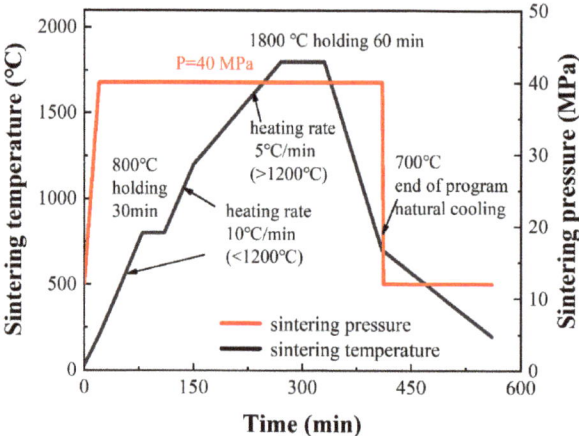

Figure 2. The schematic diagram of the sintering procedure.

2.4. Mechanical Properties

The evaluations of the mechanical properties for each group were repeated at least 5 times, and the obtained results are presented as means ± standard deviations.

2.4.1. Relative Density

Since the compactness of the as-sintered ceramics is closely related to their mechanical properties, thus, the relative density test and the corresponding results and discussion are included in the mechanical properties evaluation section. The bulk density was measured by an analytical balance (ME204, Mettler, Zurich, Switzerland) separately in air and deionized water at least 5 times, and the value of relative density (%) was calculated by the Archimedes method [11].

2.4.2. Vickers Hardness

The Vickers hardness (HV, GPa) of SiC composite ceramics with different content of Y_2O_3 was tested by a Vickers hardness tester (HV-50Z, Huayin, Laizhou, China) with a load of 20 kgf for 15 s. The length of the diagonal indentation ($2a$) and the final Vickers hardness value were obtained by the standard GB/T 4340.1-2009.

2.4.3. Fracture Toughness

The indentation fracture toughness (K_{IC}, MPa · m$^{1/2}$) of each group was obtained by the Vickers hardness tests mentioned above. The parameters, including the length of the diagonal hardness indentation ($2a$) and the crack along the edge of the indentation ($2c$), were precisely measured by the metallomicroscope (Axio Imager A2m, Zeiss, Oberkochen, Germany). Young's modulus (E) of the SiC matrix as 400 GPa was reported by Guo et al. [38]. The value of indentation fracture toughness was calculated by Equation (1) according to the Niihara [9,38,39].

$$K_{IC} = 0.0181 E^{0.4} HV^{0.6} a(c-a)^{-0.5} \tag{1}$$

2.4.4. Flexural Strength

The flexural strength (σ, MPa) was investigated by a universal testing machine (CMT5305, MTS, Eden Prairie, MN, USA) with a cross-head speed of 0.2 mm/min and a span of 14.5 mm. The maximum applied load of the flexural fracture was recorded and calculated according to the 3-point bending method to obtain the value of flexural strength. Besides, the bar-shaped samples for the flexural strength test should conform to the requirement given by the standard GB/T 3851-2015 (a size of 5.5 mm × 6.25 mm × 20 mm).

2.5. Tribological Behavior

A tribometer (MFT-5000, RTEC) was used to investigate the tribological behavior of the SiC/SiC$_w$-Y$_2$O$_3$ composite ceramics with different Y$_2$O$_3$ contents. The specimens were polished to surface roughness below 0.4 μm before the tribological test, aiming to reduce the imperfections like scratches that would interfere with the results. The linear-reciprocating wear tests were conducted at room temperature and dry conditions with a frequency of 2 Hz, P is the applied load (20 N), sliding time of 20 min, a stroke length of 5 mm, and s is the friction stroke length (12 m). YG6 balls (6 wt.% Co, Vickers hardness: 20 GPa) with a diameter of 6.35 mm were used as friction pairs in the tests. The tribological test for each group was carried out 3 times, and the representative results were exhibited in Section 3. The coefficient of friction (COF) variation with sliding time was recorded automatically by the tribometer. The partial roughness of the worn surface can be represented by the 2D profiles and 3D morphologies characterization of the wear tracks instead of measuring the entire surface roughness after the wear test. The 2D and 3D profiles of the wear tracks were reconstructed by the equipped white-light interferometric profilometer. SEM images of the wear scar were obtained to analyze the wear morphology and type. The geometric parameters of wear scars (mm) and wear volume (V, mm^3) were obtained by white light interferometer data, and the specific wear rate (WR, mm^3/(N · m)) was calculated according to Equation (2) that researchers mentioned [8,40].

$$WR = \frac{V}{Ps} \qquad (2)$$

3. Results and Discussion

Figure 3 shows the fracture morphologies of SiC/SiC$_w$-Y$_2$O$_3$ composite ceramics obtained by the three-point bending tests. The porous microstructure can be observed in the Y000 and Y025 groups (Figure 3a,b), indicating their low compactness. As Y$_2$O$_3$ content increases, the number of pores on the fracture morphologies of the Y050, Y075, and Y100 groups is significantly reduced, and the connection between the grains becomes tight (Figure 3c–e). Thus, it can be concluded that the increased Y$_2$O$_3$ content can facilitate the compacting of SiC/SiC$_w$-Y$_2$O$_3$ composite ceramics. Besides, the embedded SiC$_w$ and their pullout locations (yellow arrow) in the composite ceramics have also been pointed out. The SiC$_w$ with high elastic modulus and low crystal defects can withstand the external force and not easy to generate the whisker fracture. The matrix fracture occurs when a high external force is applied and beyond the load-bearing capacity of the SiC matrix and SiC$_w$. During fracturing, when the external force exceeds the interface bonding force between SiC$_w$ and SiC matrix, the SiC$_w$ pullout phenomenon occurs [8,9].

To further reveal the microstructure characteristic of the compacted ceramic, the morphologies of Y075 and Y100 groups after careful polishing were observed by SEM (PDBSE mode), as shown in Figure 4a,b. The quantitative EDS was performed to estimate the Y$_2$O$_3$ content of the detected points with different grayscale areas, as shown in Figure 4b,c (for Y075) and Figure 4e,f (for Y100). It can be found that Y$_2$O$_3$ (area with lighter grayscale) filled in the gaps between SiC and SiC$_w$, proving that liquefied Y$_2$O$_3$ can promote particle rearrangement and mass transfer to increase the density during sintering [12,25,27]. In the Y075 group, Y$_2$O$_3$ is evenly distributed and presents a slender outline between the SiC and SiC$_w$. Sharma et al. [40] mentioned that the glassy phase distributed between the grain boundaries could enhance the fracture toughness by combining intergranular crack mode and energy-dissipating processes in the crack wake. In contrast, Y$_2$O$_3$ in the Y100 group shows severe aggregation. It can be predicted that excess Y$_2$O$_3$ aggregated around SiC and SiC$_w$ may deteriorate the mechanical and relevant properties of the composite ceramics [26,27].

Figure 3. The fracture morphologies of SiC/SiC$_w$-Y$_2$O$_3$ composite ceramics with different Y$_2$O$_3$ contents. (**a**) Y000; (**b**) Y025; (**c**) Y050; (**d**) Y075; (**e**) Y100. The yellow arrow indicates SiC$_w$ or its location after being pullout.

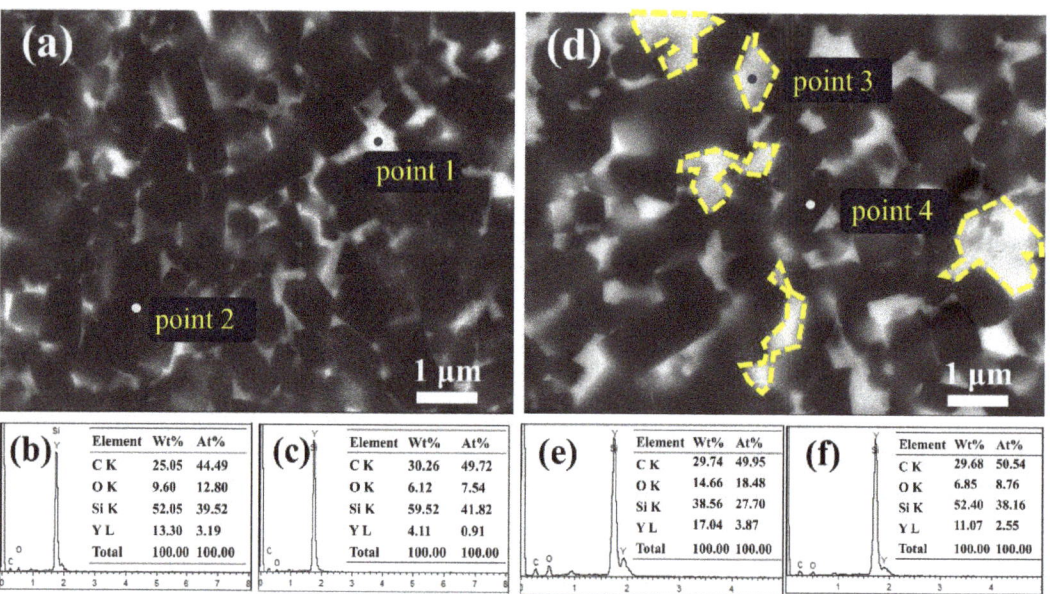

Figure 4. The PDBSE images and EDS point scan results of SiC/SiC$_w$-Y$_2$O$_3$ composite ceramics. (**a**–**c**) Y075, (**d**–**f**) Y100. Points 1 and 3 indicate the lighter grayscale area, and points 2 and 4 indicate the darker grayscale area on the polished morphologies.

XRD results are used to identify the phase composition of the SiC/SiC$_w$-Y$_2$O$_3$ composite ceramics, as shown in Figure 5. Most of the characteristic peaks of SiC can be detected (regardless of their specific crystalline structure) in all groups. The results in Figures 1 and 4 confirm the existence of SiC and SiC$_w$ in the composite ceramics. Except for Y000, the characteristic peaks (29.2° and 48.5°) belonging to PDF-#65-3178 can also be found in the rest groups, indicating the existence of Y_2O_3 in the composite ceramics. Moreover, the corresponding peak intensity of Y_2O_3 shows an increased trend with Y_2O_3 content increase. No new peaks and peaks shifting appear in the diffraction patterns, demonstrating the good stability of the three components during sintering.

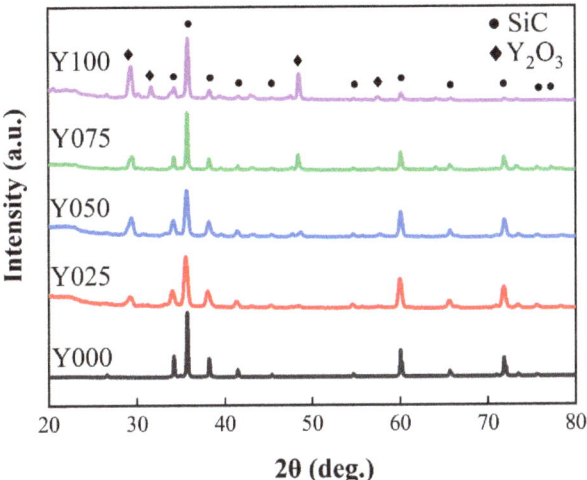

Figure 5. XRD patterns of SiC/SiC$_w$-Y_2O_3 composite ceramics with different Y_2O_3 contents.

The relative density, Vickers hardness, fracture toughness, and flexural strength of SiC/SiC$_w$-Y_2O_3 composite ceramics are shown in Figure 6. Figure 6a shows that the relative density of composite ceramic increases with Y_2O_3 content; Y000 has the lowest value of 54.3%. With Y_2O_3 content increase to 2.5 vol.%, 5 vol.%, 7.5 vol.%, and 10 vol.%, the relative densities reach to 87.6%, 91.5%, 97.0%, and 99.9%, respectively. The relative density results are consistent with the fracture morphologies shown in Figure 3, confirming that the addition of Y_2O_3 contributes to the compactness of the SiC/SiC$_w$-Y_2O_3 composite ceramics. It's worth noting that the poor compactness of Y000 reveals the formula failed when sintering in this process condition; therefore, many valid mechanical and tribological data cannot be obtained. For example, although the Vickers hardness of Y000 was obtained, it was only 1.3 GPa and at least an order of magnitude smaller than the other groups. Hence, the Y000 group was abandoned in subsequent tests. Compare the Y000 specimen with the other 4 groups containing Y_2O_3, confirming that Y_2O_3 additives have a noticeable promoting effect on the microstructure and mechanical properties of SiC composite ceramics. With the Y_2O_3 content increasing from 2.5 vol.% to 7.5 vol.%, the Vickers hardness increases from 10.9 GPa to 21.6 GPa, then decreases to 18.9 GPa when Y_2O_3 content is 10 vol.%. The results indicate that suitable Y_2O_3 content ($\leq$7.5 vol.%) can increase the Vickers hardness, and excessive Y_2O_3 content (10 vol.%) can weaken the Vickers hardness. Similar variations can be found in the fracture toughness (Figure 6c) and flexural strength results (Figure 6d). The values increase when Y_2O_3 content increases from 2.5 vol.% to 7.5 vol.% and then decrease when Y_2O_3 content reaches 10 vol.%. Of course, although the tested mechanical properties of Y100 are not the best, it is still slightly better than Y050 and Y025, proving that excessive Y_2O_3 does not significantly deteriorate the relevant performance dramatically. The hampered mechanical properties of Y100 are strongly affected by their microstructure since the aggregated Y_2O_3 can decrease its load-bearing capacity attributed to the poorer mechanical properties of Y_2O_3 compared with SiC [37]. Besides, the excessive Y_2O_3 made redundant liquid phase connect in a larger potential flaw fracture origin [26].

Seo et al. [19] reported the SiC-2 vol.%Y_2O_3 ceramic which was fabricated by hot-press sintering at 2000 °C, 40 MPa, N_2 for 3 h. The relative density, fracture toughness, and flexural strength are 99.1%, 3.4 MPa · m$^{1/2}$, and 586 MPa, respectively. SiC-2 vol.%Y$_2$O$_3$ ceramic shows slightly higher relative density and flexural strength than SiC/SiC$_w$-7.5 vol.%Y$_2$O$_3$ composite ceramic (Y075) sintered at 1800 °C in this study. Both studies indicate that Y$_2$O$_3$ can improve the densification of the as-sintered ceramics, and the SiC composite ceramics with an appropriate amount of Y$_2$O$_3$ can reduce the hot-pressed sintering temperature to a comparatively lower one (1800 °C). Furthermore, compared with Y075, the fracture toughness of SiC-2 vol.%Y$_2$O$_3$ ceramics decreased by 55.8%. It can be speculated that the SiC$_w$ reinforced [6–9] and the Y$_2$O$_3$ pinned in [25] can both improve the mechanical properties, particularly the fracture toughness of the SiC composite ceramics. It can be concluded that incorporated Y$_2$O$_3$ and SiC$_w$ can work together to improve the compactness of microstructure and the mechanical performance of the SiC composite ceramics.

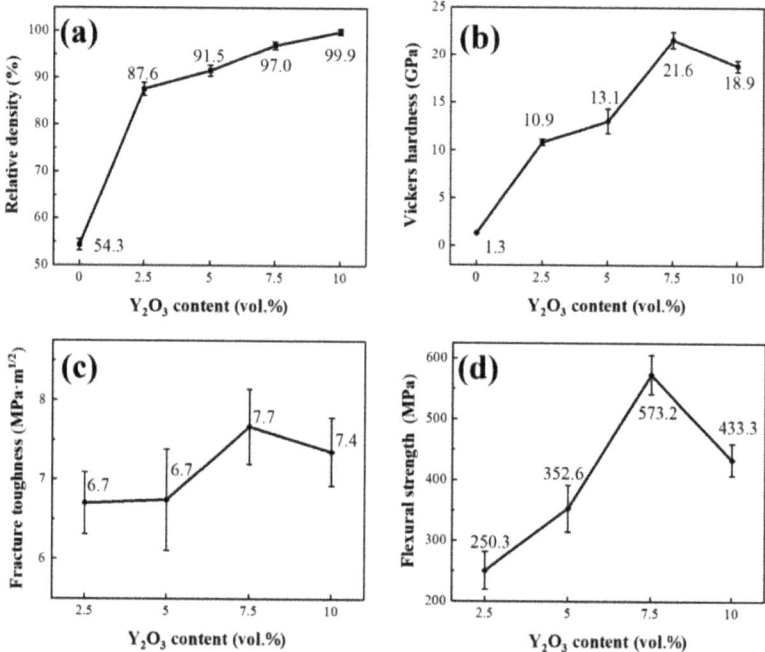

Figure 6. The variations of (**a**) relative density, (**b**) Vickers hardness, (**c**) fracture toughness, and (**d**) flexural strength of SiC/SiC$_w$-Y$_2$O$_3$ composite ceramics.

The COF variations of SiC/SiC$_w$-Y$_2$O$_3$ composite ceramics with sliding time are displayed in Figure 7a. In all groups, the COF sharply rises at the initial stage, then decreases and shows a relatively stable status after sliding for ~200 s. After contact with each other, the asperities on SiC/SiC$_w$-Y$_2$O$_3$ composite ceramic and YG6 ball increase the contact stress between them so that a considerable COF value can be obtained. After the run-in stage, the relatively stable COF indicates that the contact surface of the specimen and ball increases, indicating the decreased surface roughness of the grinding bodies. Although the COF at the stable stage does not change dramatically, slight increases of COF in groups with lower contents (Y025 and Y050) can still be seen. The average COF of Y025, Y050, Y075, and Y100 during the sliding period are 0.36, 0.35, 0.34, and 0.32, respectively, indicating that incorporating Y$_2$O$_3$ can decrease the COF of SiC/SiC$_w$-Y$_2$O$_3$ composite ceramics. Combined with the formula of SiC/SiC$_w$-Y$_2$O$_3$ composite ceramics, it can be boldly speculated that the worn Y$_2$O$_3$ debris can more easily fill the pits of the wear scars than SiC and SiC$_w$, reduce the roughness, and thus slightly reduce the COF. The COF and antifriction effect of Y$_2$O$_3$ is positively correlated with its content [37,40–42].

Figure 7b depicts the wear track profiles of SiC/SiC$_w$-Y$_2$O$_3$ composite ceramics. It can be seen both the width and depth of the wear track of composite ceramics decrease in groups with Y$_2$O$_3$ content ≤7.5 vol.%. As the Y$_2$O$_3$ content continues to increase, the width and depth of the wear track of Y100 increase, but it is still smaller than Y050. Besides, the wear scar profiles of Y075 and Y100 are smoother than Y025 and Y050, indicating the more compact microstructure and better mechanical properties of the latter groups. Figure 7c1–c4 vividly displays the 3D morphology of the wear track on each sample surface. These images exhibit the depth and width of the wear track (consistent with Figure 7b) and can reveal the surface roughness inside the wear track. As can be seen, the wear tracks of Y025 and Y050 possess larger roughness than Y100 and Y075, and the Y075 is the smallest. Generally, in the case of the same materials type, the smoother the surface, the lower the friction coefficient, and the better the wear resistance [43,44].

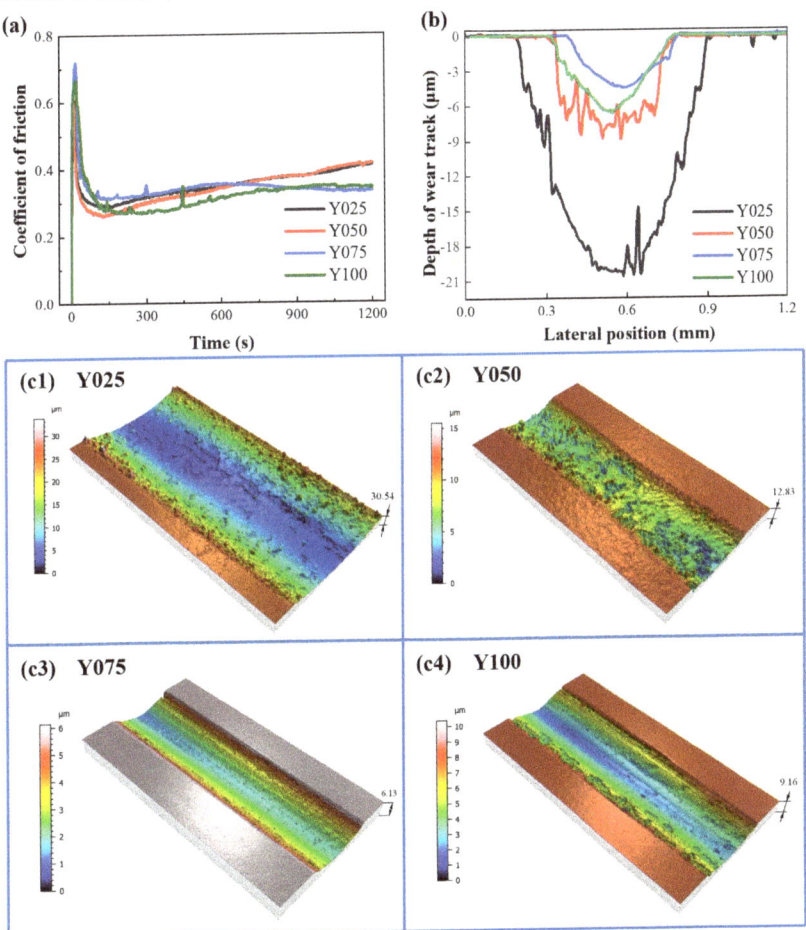

Figure 7. (**a**) The coefficient of friction, (**b**) the 2D, and (**c1–c4**) the 3D wear track profile of SiC/SiC$_w$-Y$_2$O$_3$ composite ceramics after grinding with YG6 balls at 20 N for 20 min: (**c1**) Y025; (**c2**) Y050, (**c3**) Y075, (**c4**) Y100.

The maximum depth, width, wear volume, and specific wear rate of the wear track of SiC/SiC$_w$-Y$_2$O$_3$ composite ceramics are listed in Table 1. The maximum depth and width are consistent with the results of 2D and 3D profiles in Figure 7b,c1–c4. Moreover, with Y$_2$O$_3$ content increasing from 2.5 vol.% to 7.5 vol.%, the wear volume and specific wear rate reach the lowest value of 3.82×10^{-3} mm^3 and 1.59×10^{-5} mm^3/(N · m). As the Y$_2$O$_3$ content continues to increase, the wear volume and specific wear rate slightly increase but are still lower than the Y050 group. Thus, it can be concluded that high Y$_2$O$_3$ addition can reduce the COF, and Y$_2$O$_3$ additives can significantly improve the wear resistance of SiC/SiC$_w$-Y$_2$O$_3$ composite ceramics. SiC composite ceramics with 7.5 vol.%Y$_2$O$_3$ shows the best wear resistance than others.

Table 1. The maximum depth, maximum width, wear volume, and specific wear rate of the wear track of SiC/SiC$_w$-Y$_2$O$_3$ composite ceramics with different Y$_2$O$_3$ contents.

	Maximum Depth (mm)	Maximum Width (mm)	Wear Volume (mm^3)	Specific Wear Rate (mm^3/(N · m))
Y025	2.63×10^{-2}	7.30×10^{-1}	3.25×10^{-2}	1.35×10^{-4}
Y050	1.25×10^{-2}	5.01×10^{-1}	7.74×10^{-3}	3.23×10^{-5}
Y075	7.44×10^{-3}	4.09×10^{-1}	3.82×10^{-3}	1.59×10^{-5}
Y100	8.23×10^{-3}	4.51×10^{-1}	6.46×10^{-3}	2.69×10^{-5}

SEM images of wear scars reveal the correlation between the microstructure and wear resistance. As shown in Figure 8, the surface morphologies of wear scars (first line) are in accordance with the 3D profiles shown in Figure 7c1–c4. As for Y025, flaky debris does not connect tight in the microscale covered on the wear scar surface. This morphology was caused by the cold-welded debris onto the edge of the SiC. The flaky debris filled the holes of the hot-pressed ceramic, but due to the low friction heat and insufficient quantity, they do not have tight bonding with the substance below [43]. Meanwhile, the asperities bear greater friction and lead to worse wear resistance. The wear morphology and mechanism of Y050 are similar to that of Y025, but the wear surface of Y050 is mildly flat, and the wear resistance is improved over Y025 due to the increased relative density. Adding Y$_2$O$_3$ up to 7.5 vol.% and 10 vol.% further improves the relative density of composite ceramics, but excessive Y$_2$O$_3$ aggregated around SiC and SiC$_w$ (especially Y100) weakens the Vickers hardness and the subsequent wear resistance [27,45]. As seen in Figure 8c2,d2, no sintering holes exist in Y075 and Y100, but the size of cracks in Y100 is larger than in Y075, indicating poor mechanical properties of Y100. The lower hardness of Y$_2$O$_3$ than SiC is also responsible for the decreased wear resistance of Y100 [37]. Y075 group possesses the optimal wear resistance, and the abrasive wear plays an important role, as evidenced by smooth wear scar surfaces and long and thin furrows. The specific wear rate of Y075 is 11.8% than Y025.

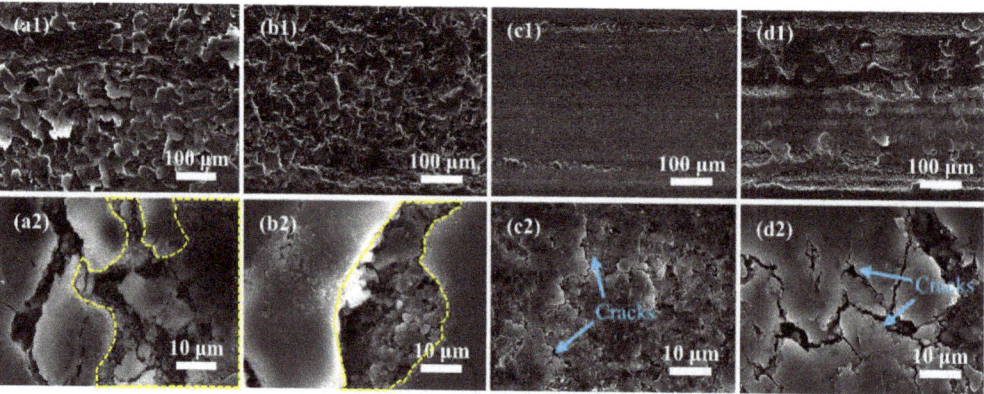

Figure 8. The SEM morphologies of the wear track of SiC/SiCw-Y$_2$O$_3$ composite ceramics with different Y$_2$O$_3$ contents. (**a1,a2**) Y025; (**b1,b2**) Y050; (**c1,c2**) Y075; (**d1,d2**) Y100. The yellow-dotted polygon outlines the sintering holes filled with wear debris.

4. Conclusions

SiC/SiC$_w$-Y$_2$O$_3$ composite ceramics with 10 vol.% SiC$_w$ and different Y$_2$O$_3$ contents were fabricated by hot-pressed sintering at 1800 °C. The effects of Y$_2$O$_3$ content on the microstructure, mechanical properties, and tribological behaviors were systematically investigated. It is found that compact SiC/SiC$_w$ cannot be hot-pressed without Y$_2$O$_3$ additives at 1800 °C and 40 MPa. With Y$_2$O$_3$ content increase, the microstructures are getting compact with increased relative density, but excessive Y$_2$O$_3$ leads to its aggregation. Moreover, with Y$_2$O$_3$ content increasing from 2.5 vol.% to 7.5 vol.%, the Vickers hardness, fracture toughness, and flexural strength all increase to the optimal values. As Y$_2$O$_3$ content increased to 10 vol.%, those properties were slightly reduced. Due to the compact microstructure, proper Y$_2$O$_3$ content, and excellent mechanical properties, SiC/SiC$_w$-7.5 vol.%Y$_2$O$_3$ ceramic (Y075) shows optimal wear resistance. The specific wear rate of Y075 is 11.8% of that for Y025 (SiC/SiC$_w$-2.5 vol.%Y$_2$O$_3$ ceramic). The excellent mechanical properties and good wear resistance of Y075 highlight its potential future application.

Author Contributions: Conceptualization, J.W. and L.Z.; methodology, S.Z. and J.W.; software, S.Z. and L.D.; validation, S.Z. and H.C.; formal analysis, X.L. and F.W.; investigation, Z.T., L.S. and S.C.; resources, L.Z.; data curation: S.Z.; data interpretation: S.Z. and S.D. writing—original draft preparation, M.Z., J.W. and S.Z; writing—review and editing, M.Z., J.W. and L.Z.; visualization, N.Y.; supervision, S.D.; project administration, L.Z. and S.C.; funding acquisition, L.S., N.Y. and L.Z. All authors have read and agreed to the published version of the manuscript.

Funding: The authors would like to acknowledge the National Natural Science Foundation of China (No. 21671145 and No. 52201187), the Zhejiang Province key research and development plan project (2023C01082), the General Scientific Research Project of Zhejiang Provincial Education Department (No. Y202249336), the Natural Science Foundation of Zhejiang Province (LQ19E050004), the Agricultural Science and Technology Key R & D Program of Taizhou (NYJBGS202201).

Institutional Review Board Statement: This article does not need ethical approval.

Informed Consent Statement: This article does not involve human experiments.

Data Availability Statement: Data are available from the corresponding author on reasonable request.

Conflicts of Interest: The authors declare no conflict of interest.

References

1. Xu, M.; Girish, Y.R.; Rakesh, K.P.; Wu, P.; Manukumar, H.M.; Byrappa, S.M.; Udayabhanu; Byrappa, K. Recent advances and challenges in silicon carbide (SiC) ceramic nanoarchitectures and their applications. *Mater. Today Commun.* **2021**, *28*, 102533. [CrossRef]
2. Padture, N.P. Advanced structural ceramics in aerospace propulsion. *Nat. Mater.* **2016**, *15*, 804–809. [CrossRef] [PubMed]
3. Kang, S.-J.L. What we should consider for full densification when sintering. *Materials* **2020**, *13*, 3578. [CrossRef]
4. Shimoda, K.; Hinoki, T. Effects of fiber volume fraction on the densification and mechanical properties of unidirectional SiCf/SiC-matrix composites. *J. Eur. Ceram. Soc.* **2021**, *41*, 1163–1170. [CrossRef]
5. Wei, C.; Zhang, Z.; Ma, X.; Liu, L.; Wu, Y.; Li, D.; Wang, P.; Duan, X. Mechanical and ablation properties of laminated ZrB$_2$-SiC ceramics with Si3N4 whisker interface. *Corros. Sci.* **2022**, *10*, 17003–17009. [CrossRef]
6. Li, S.; Zhang, Y.; Han, J.; Zhou, Y. Fabrication and characterization of SiC whisker reinforced reaction bonded SiC composite. *Ceram. Int.* **2013**, *39*, 449–455. [CrossRef]
7. Song, N.; Zhang, H.-B.; Liu, H.; Fang, J.-Z. Effects of SiC whiskers on the mechanical properties and microstructure of SiC ceramics by reactive sintering. *Ceram. Int.* **2017**, *43*, 6786–6790. [CrossRef]
8. Yang, Y.; Zhu, T.; Sun, N.; Liang, X.; Li, Y.; Wang, H.; Xie, Z.; Sang, S.; Dai, J. Mechanical and tribological properties of SiC whiskers reinforced SiC composites via oscillatory pressure sintering. *Int. J. Appl. Ceram. Technol.* **2023**, *12*, 1–12. [CrossRef]
9. Zhang, L.; Yang, H.; Guo, X.; Shen, J.; Zhu, X. Preparation and properties of silicon carbide ceramics enhanced by TiN nanoparticles and SiC whiskers. *Scr. Mater.* **2011**, *65*, 186–189. [CrossRef]
10. Kim, K.-J.; Lim, K.-Y.; Kim, Y.-W. Influence of Y$_2$O$_3$ addition on electrical properties of β-SiC ceramics sintered in nitrogen atmosphere. *J. Eur. Ceram. Soc.* **2012**, *32*, 4401–4406. [CrossRef]
11. Xie, W.; Fu, Q.; Cheng, C. Oxidation behavior of different La$_2$O$_3$-content modified SiC ceramic at 1700 degrees C. *Ceram. Int.* **2021**, *47*, 11560–11567. [CrossRef]
12. Tabata, S.; Hirata, Y.; Sameshima, S.; Matsunaga, N.; Ijichi, K. Liquid phase sintering and mechanical properties of SiC with rare-earth oxide. *J. Ceram. Soc. Jpn.* **2006**, *114*, 247–252. [CrossRef]

13. Liang, H.; Yao, X.; Zhang, J.; Liu, X.; Huang, Z. The effect of rare earth oxides on the pressureless liquid phase sintering of alpha-SiC. *J. Eur. Ceram. Soc.* **2014**, *34*, 2865–2874. [CrossRef]
14. Raju, K.; Yoon, D.-H. Sintering additives for SiC based on the reactivity: A review. *Ceram. Int.* **2016**, *42*, 17947–17962. [CrossRef]
15. Maity, T.; Kim, Y.-W. High-temperature strength of liquid-phase-sintered silicon carbide ceramics: A review. *Int. J. Appl. Ceram. Technol.* **2022**, *19*, 130–148. [CrossRef]
16. Strecker, K.; Ribeiro, S.; Oberacker, R.; Hoffmann, M.J. Influence of microstructural variation on fracture toughness of LPS-SiC ceramics. *Int. J. Refract. Met. Hard Mater.* **2004**, *22*, 169–175. [CrossRef]
17. Santos, A.; Ribeiro, S. Liquid phase sintering and characterization of SiC ceramics. *Ceram. Int.* **2018**, *44*, 11048–11059. [CrossRef]
18. Ge, S.; Yao, X.; Liu, Y.; Duan, H.; Huang, Z.; Liu, X. Effect of Sintering Temperature on the Properties of Highly Electrical Resistive SiC Ceramics as a Function of Y_2O_3-Er_2O_3 Additions. *Materials* **2020**, *13*, 4768. [CrossRef]
19. Seo, Y.-K.; Kim, Y.-W.; Kim, K.-J. Electrically conductive SiC-BN composites. *J. Eur. Ceram. Soc.* **2016**, *36*, 3879–3887. [CrossRef]
20. Zhu, Y.; Qin, Z.; Chai, J.; Shen, T.; Zhou, Y.; Wang, Z. Effects of sintering additives on microstructure and mechanical properties of hot-pressed alpha-SiC. *Metall. Mater. Trans. A* **2022**, *53*, 1188–1199. [CrossRef]
21. Lim, K.-Y.; Kim, Y.-W.; Kim, K.J. Mechanical properties of electrically conductive silicon carbide ceramics. *Ceram. Int.* **2014**, *40*, 10577–10582. [CrossRef]
22. Kueck, A.; Ramasse, Q.; De Jonghe, L.; Ritchie, R. Atomic-scale imaging and the effect of yttrium on the fracture toughness of silicon carbide ceramics. *Acta Mater.* **2010**, *58*, 2999–3005. [CrossRef]
23. Gao, J.; Xiao, H.; Du, H. Effect of Y_2O_3 addition on ammono sol–gel synthesis and sintering of Si_3N_4–SiC nanocomposite powder. *Ceram. Int.* **2003**, *29*, 655–661. [CrossRef]
24. Zhang, Y.; Yu, X.; Gu, H.; Yao, D.; Zuo, K.; Xia, Y.; Yin, J.; Liang, H.; Zeng, Y.-P. Microstructure evolution and high-temperature mechanical properties of porous Si_3N_4 ceramics prepared by SHS with a small amount of Y_2O_3 addition. *Ceram. Int.* **2021**, *47*, 5656–5662. [CrossRef]
25. Cheng, L.; Xie, Z.; Liu, J.; Wu, H.; Jiang, Q.; Wu, S. Effects of Y_2O_3 on the densification and fracture toughness of SPS-sintered TiC. *Mater. Res. Innov.* **2018**, *22*, 7–12. [CrossRef]
26. Zhang, X.; Li, X.; Han, J.; Han, W.; Hong, C. Effects of Y_2O_3 on microstructure and mechanical properties of ZrB_2–SiC ceramics. *J. Alloys Compd.* **2008**, *465*, 506–511. [CrossRef]
27. Kováčová, Z.; Bača, Ľ.; Neubauer, E.; Kitzmantel, M. Influence of sintering temperature, SiC particle size and Y_2O_3 addition on the densification, microstructure and oxidation resistance of ZrB_2-SiC ceramics. *J. Eur. Ceram. Soc.* **2016**, *36*, 3041–3049. [CrossRef]
28. Aguirre, T.G.; Lamm, B.W.; Cramer, C.L.; Mitchell, D.J. Zirconium-diboride silicon-carbide composites: A review. *Ceram. Int.* **2022**, *48*, 7344–7361. [CrossRef]
29. Aguirre, T.G.; Cramer, C.L.; Cakmak, E.; Lance, M.J.; Lowden, R.A. Processing and microstructure of ZrB_2–SiC composite prepared by reactive spark plasma sintering. *Results Mater.* **2021**, *11*, 100217. [CrossRef]
30. Li, W.; Zhang, X.; Hong, C.; Han, J.; Han, W. Hot-pressed ZrB_2-SiC-YSZ composites with various yttria content: Microstructure and mechanical properties. *Mater. Sci. Eng. A* **2008**, *494*, 147–152. [CrossRef]
31. Serbenyuk, T.B.; Prikhna, T.O.; Sverdun, V.B.; Sverdun, N.V.; Moshchil', V.Y.; Ostash, O.P.; Vasyliv, B.D.; Podhurska, V.Y.; Kovylyaev, V.V.; Chasnyk, V.I. Effect of the additive of Y_2O_3 on the structure formation and properties of composite materials based on AlN-SiC. *J. Superhard Mater.* **2018**, *40*, 8–15. [CrossRef]
32. Wang, J.; Zuo, D.; Zhu, L.; Li, W.; Tu, Z.; Dai, S. Effects and influence of Y_2O_3 addition on the microstructure and mechanical properties of binderless tungsten carbide fabricated by spark plasma sintering. *Int. J. Refract. Met. Hard Mater.* **2018**, *71*, 167–174. [CrossRef]
33. Wang, J.; Zuo, D.; Zhu, L.; Tu, Z.; Lin, X.; Wu, Y.; Li, W.; Zhang, X. Effect of Y_2O_3 addition on high-temperature oxidation of binderless tungsten carbide. *Front. Mater.* **2021**, *8*, 645612. [CrossRef]
34. Liu, Y.; Li, X.; Zhou, J.; Fu, K.; Wei, W.; Du, M.; Zhao, X. Effects of Y_2O_3 addition on microstructures and mechanical properties of WC-Co functionally graded cemented carbides. *Int. J. Refract. Met. Hard Mater.* **2015**, *50*, 53–58. [CrossRef]
35. Hu, C.; Xiang, W.; Chen, P.; Li, Q.; Xiang, R.; Zhou, L. Influence of Y_2O_3 on densification, flexural strength and heat shock resistance of cordierite-based composite ceramics. *Ceram. Int.* **2022**, *48*, 74–81. [CrossRef]
36. Barick, P.; Chakravarty, D.; Saha, B.P.; Mitra, R.; Joshi, S.V. Effect of pressure and temperature on densification, microstructure and mechanical properties of spark plasma sintered silicon carbide processed with beta-silicon carbide nanopowder and sintering additives. *Ceram. Int.* **2016**, *42*, 3836–3848. [CrossRef]
37. Gupta, S.; Sharma, S.K.; Kumar, B.V.M.; Kim, Y.-W. Tribological characteristics of SiC ceramics sintered with a small amount of yttria. *Ceram. Int.* **2015**, *41*, 14780–14789. [CrossRef]
38. Guo, X.; Yang, H.; Zhang, L.; Zhu, X. Sintering behavior, microstructure and mechanical properties of silicon carbide ceramics containing different nano-TiN additive. *Ceram. Int.* **2010**, *36*, 161–165. [CrossRef]
39. Niihara, K.; Morena, R.; Hasselman, D.P.H. Evaluation of K_{IC} of brittle solids by the indentation method with low crack-to-indent ratios. *J. Mater. Sci. Lett.* **1982**, *1*, 533–538. [CrossRef]
40. Sharma, S.K.; Kumar, B.V.M.; Kim, Y.-W. Tribological behavior of silicon carbide ceramics—A Review. *J. Korean Ceram. Soc.* **2016**, *53*, 581–596. [CrossRef]
41. Kumar, B.V.M.; Kim, Y.-W.; Lim, D.-S.; Seo, W.-S. Influence of small amount of sintering additives on unlubricated sliding wear properties of SiC ceramics. *Ceram. Int.* **2011**, *37*, 3599–3608. [CrossRef]

42. Zhang, W.; Yamashita, S.; Kita, H. Progress in tribological research of SiC ceramics in unlubricated sliding-A review. *Mater. Des.* **2020**, *190*, 108528. [CrossRef]
43. Fischer, T.; Zhu, Z.; Kim, H.; Shin, D. Genesis and role of wear debris in sliding wear of ceramics. *Wear* **2000**, *245*, 53–60. [CrossRef]
44. Xia, H.; Qiao, G.; Zhou, S.; Wang, J. Reciprocating friction and wear behavior of reaction-formed SiC ceramic against bearing steel ball. *Wear* **2013**, *303*, 276–285. [CrossRef]
45. Cho, S.-J.; Um, C.-D.; Kim, S.-S. Wear and Wear Transition in Silicon Carbide Ceramics during Sliding. *J. Am. Ceram. Soc.* **1996**, *79*, 1247–1251. [CrossRef]

Disclaimer/Publisher's Note: The statements, opinions and data contained in all publications are solely those of the individual author(s) and contributor(s) and not of MDPI and/or the editor(s). MDPI and/or the editor(s) disclaim responsibility for any injury to people or property resulting from any ideas, methods, instructions or products referred to in the content.

Article

Interfacial Microstructure and Shear Behavior of the Copper/Q235 Steel/Copper Block Fabricated by Explosive Welding

Jiansheng Li [1], Zuyuan Xu [1], Yu Zhao [1,*], Wei Jiang [1], Wenbo Qin [2], Qingzhong Mao [3], Yong Wei [1] and Banglun Wang [1,*]

1. School of Materials Science and Engineering, Anhui Polytechnic University, Wuhu 241000, China
2. School of Engineering and Technology, China University of Geosciences (Beijing), Beijing 100083, China
3. School of Materials Science and Engineering, Nanjing University of Science and Technology, Nanjing 210094, China
* Correspondence: zhaoyu@ahpu.edu.cn or zhaoyu0816@126.com (Y.Z.); banglun@foxmail.com (B.W.)

Abstract: A copper/Q235 steel/copper composite block with excellent bonding interfaces was prepared by explosive welding which was a promising technique to fabricate laminates. The microstructure and mechanical properties of the interfaces were investigated via the tensile-shear test, optical microscope (OM), X-ray diffraction (XRD), scanning electron microscope (SEM), and electron back-scattered diffraction (EBSD). The results showed that the shear strength of the upper-interface and lower-interfaces of the welded copper/steel are higher than ~235 MPa and ~222 MPa, respectively. The specimens failed fully within the copper and not at the bonding interface. It was attributed to: (1) no cavities and cracks at the interface; (2) the interface formed a metallurgical bonding including numerous ultra-fine grains (UFGs) which can significantly improve the plastic deformation coordination at the interface and inhibit the generation of micro-cracks.

Keywords: copper/steel/copper composite; explosive welding; microstructure; shear behavior; interface

Citation: Li, J.; Xu, Z.; Zhao, Y.; Jiang, W.; Qin, W.; Mao, Q.; Wei, Y.; Wang, B. Interfacial Microstructure and Shear Behavior of the Copper/Q235 Steel/Copper Block Fabricated by Explosive Welding. *Coatings* **2023**, *13*, 600. https://doi.org/10.3390/coatings13030600

Academic Editors: Alina Vladescu and Ajay Vikram Singh

Received: 4 February 2023
Revised: 2 March 2023
Accepted: 9 March 2023
Published: 11 March 2023

Copyright: © 2023 by the authors. Licensee MDPI, Basel, Switzerland. This article is an open access article distributed under the terms and conditions of the Creative Commons Attribution (CC BY) license (https://creativecommons.org/licenses/by/4.0/).

1. Introduction

Copper-steel composite laminates combine the excellent electrical/thermal conductivity of pure copper with the high strength and corrosion resistance of steel, and have a wide application in many fields, such as nuclear island cooling systems, vacuum chamber heat exchange systems, copper electrolytic refining equipment [1–3]. Good metallurgical bonding of copper and steel has been challenging due to the large melting point difference and low mutual solubility. Traditional copper/steel composite laminate preparation methods mainly include accumulative rolling [4], diffusion welding [5], magnetron sputtering [6], and casting [7]. Compared with the processes mentioned above, explosive welding is an effective process to weld two or more dissimilar laminates, when there are distinct differences in metal properties such as melting point, strength and so on [8]. Explosive welding is a metallurgical process that uses the shock wave generated by the explosion of explosives to subject the metal to a high-speed impact and bond it in a short period of time. Of course, there are many factors that can affect the quality of welding, such as the falling speed of the flyer plate, static angle, bending angle, burst speed of the explosive, the thickness and nature of the flyer plate and substrate [9,10].

The bonding interface has a decisive influence on the mechanical properties of the welded composites. This is because failure often occurs near the weld interface. The development of industry also places higher desire on explosive welding: (1) revealing failure mechanisms of composite materials at the welded interface; (2) improving the bonding strength of welded composite materials.

Presently, most studies on explosive welding are focused on the wave morphology, bonding strength, and microstructure at the interface of two welded dissimilar metals [10–12]. The bonding interface of explosive welding is often wavy, and the periodic fluctuation of shear stress is considered to be an important factor [13]. Gao et al. [14] investigated the effect of stand-off distance on the bonding strength and microstructure of Al/Ni sheets, and found that with the increasing of stand-off distance, the welded interface becomes more undulating, showing in turn straight, wavy and continuously melted. Zhou et al. [8] employed tensile shear test with in-situ scanning electron microscope (SEM) to investigate the interfacial bonding properties of the Q235/TA2 composite plate and micro-cracks generation and propagation in the welded interface. The results suggested that failure of the interface is the combination of the cavities, micro-cracks, and brittle intermetallic compounds. Under the tensile-shear condition, micro-cracks develop at the welded interface, associate, and subsequently develop along the interface wave direction. In order to increase the interfacial bonding of explosive welding, friction stir processing (FSP) was used as a means of optimization [11]. The FSP not only reduces defects at the interface, but introduces the formation of nano-grains during stirring, which significantly improves the bonding strength of the interface. Ismail et al. [15] found that the addition of carbon nanotubes (CNTs) between two plates increases the resistance of the solder joint to the blast wave. Based on the above studies, it can obviously be found that many efforts were devoted to exploit the interfacial microstructures and improve the interface bonding strength for the explosive welded composite with two kinds of dissimilar metals. In general, relatively little effort has been devoted to fabricate multi-layered composites via explosive welding, especially for a widely used copper-steel composite, while the microstructure and properties of the weld interfaces are worth exploring.

In this work, a three-layered copper/Q235 steel/copper block was fabricated by an explosive welding technique. The interfacial microstructure and shear behavior of the copper/Q235 steel/copper block were clearly explored and revealed.

2. Experimental

Ordinary carbon structural steel plates (Q235, C-0.125, Si-0.072, Mn-0.145, P-0.017, S-0.015, Fe-balance in wt.%) and pure copper (Fe-0.003, Si-0.028, S-0.004, Cu-balance in wt.%) plates were selected. All plates had the same dimensions in $110 \times 110 \times 3$ mm^3. The commonly used parallel type preparation was employed to prepare copper/Q235 steel/copper composite as schematically revealed in Figure 1a above and all explosions took place in a sand pool. An ammonium nitrate fuel oil (ANFO) mixture having a density of 0.90 g/cm^3 was selected as the explosive material. The explosive material with a detonation velocity of 2200 m/s and a thickness of 65 mm was placed on a Cu plate. The stand-off distance was 6 mm. Under the action of the explosion, the flyer plate near the detonation point was accelerated to collide with the lower plate at an angle and forming a welded interface, as shown in Figure 1a below. After explosive welding, the thicknesses of the three plates from top to bottom were ~2.65 mm (copper), ~2.9 mm (Q235 steel) and ~2.9 mm (copper), respectively, two interfaces were referred to as upper-interface and lower-interface.

The mechanical properties and microstructure of explosion welded composites tend to differ along the direction of explosive wave propagation, especially the position near and far from the detonation point. Therefore, the samples used in this experiment were strictly cut from the same position of the composite. All mechanical tests and microstructure observations were performed on samples cut in the direction perpendicular to explosion direction from laminates.

X-ray diffraction (XRD) measurements were carried out on a Bruker AXS D8 diffractometer with Cu Kα radiation. The 2θ angle was selected from 40° to 100° and the scanning speed was 6° min^{-1}. Electron back-scattered diffraction (EBSD) analysis was performed on a field emission SEM (Quant 250FEG, FEI, America) equipped with a fully automatic Oxford Instruments Aztec2.0 EBSD system (Channel 5 Software, Oxford Instruments, England,

Oxford). The scanning step size and accelerate voltage were 1.5 μm and 20 kV, respectively. The fracture surfaces of specimens were also investigated by this equipment.

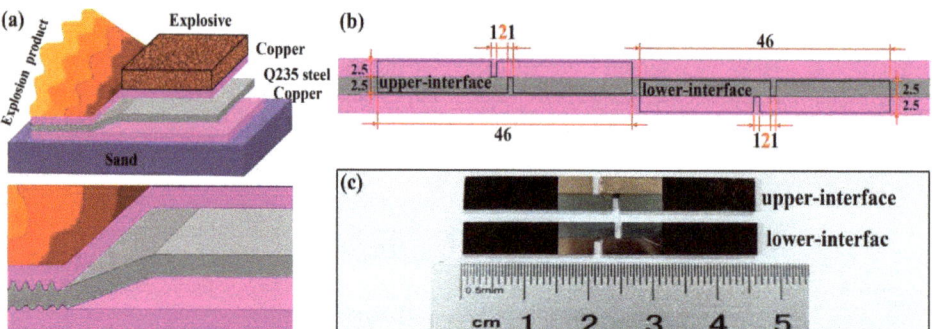

Figure 1. Schematic diagram of the experimental setup for explosive bonding process (**a**) and double notched tensile-shear specimen (**b**); the tensile-shear specimens (**c**).

The tensile-shear tests on the explosion-welded upper-interface and lower-interface were performed using an AGS-X100KN electronic universal tester (Shimadzu, Kyoto, Japan) with a movement speed of 1 mm/min. Tensile-shear specimens with a thickness of 1 mm were cut as parallel to the bonding interface from the explosively welded copper and Q235 steel plates by wire-electrode cutting and processed according to the standard of American Society for Testing and Materials (ASTM A265-03), as shown in Figure 1b,c. The specimens were notched with a width of 1 mm at two edges. It should be noticed that in this study, the form of the interface close to the explosive layer varies more significantly with the upper copper plate than with the interface far from the explosive layer. Both shear strength of the upper-interface and lower-interface of the bonding composites were considered in this work. During tensile-shear tests, the load was continuously applied to the specimen until failure fully occurred.

3. Results and Discussion

As shown in Figure 2a,b, the metallographic results of the initial microstructure indicate that pure copper has equiaxed grains with a mean size of ~17 μm, and Q235 steel (mean grain size of ~14 μm) is composed of a homogeneous mixture of ~90% ferrite (white part in Figure 2b) and ~10% pearlite (black part in Figure 2b). It is well known that pearlite is a mechanical mixture of alternating lamellar ferrite and Fe_3C [16,17]. According to the XRD results (Figure 2c), there is no other types of tissue in the matrix of both pure copper and Q235 steel. It should be noted that by comparing the PDF card (PDF#06-0696), the four diffraction peaks in the figure corresponded to the (110), (200), (211) and (220) diffraction peaks of the ferrite. This is mainly because the Fe_3C in the matrix was too small, so there was no obvious diffraction peak. The XRD results for the ND (normal direction of rolling) plane of the copper plate are shown in Figure 2c (marked in red). Compared with the XRD spectra of standard annealed copper (PDF#65-9743), it can be found that the (110) and (100) diffraction peaks at the ND plane have an increased percentage of intensity. This is attributed to the fact that copper sheets in the rolled state usually have a strong brass texture ({110}<112>) which indicates that the {110} crystalline plane parallel to the ND plane and the <112> crystal direction parallel to the RD (rolling direction). After annealing treatment, although the grains recrystallize and become equiaxed, part of the crystallographic orientation of grains is inherited [18].

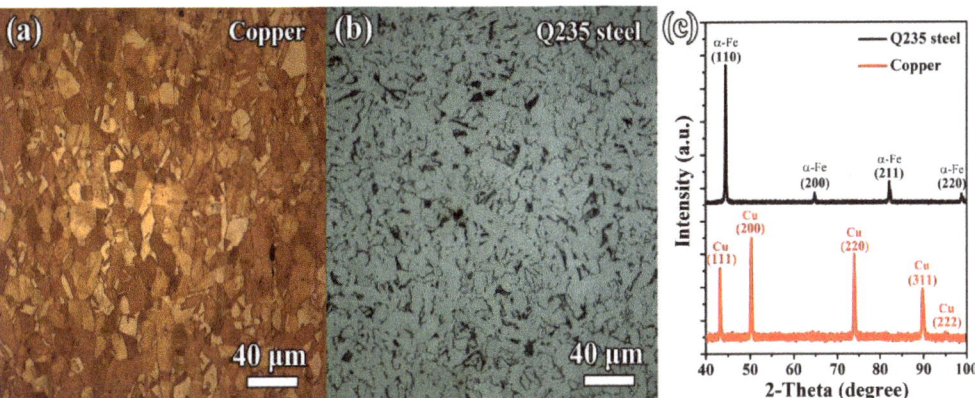

Figure 2. Optical microscope images of pure copper (**a**) and Q235 steel (**b**) in the initial state, X-ray diffraction analysis on the copper and steel plates (**c**).

Figure 3a shows the metallographic results of the welded sandwich copper/Q235 steel/copper composites which indicates that the bonding interfaces have high quality joining without defects in forms of pores and micro-cracks. As with other other explosively dissimilar welded joints [2,19,20], upper-interface and lower-interface both have wavy morphology. When the explosion energy is certain, with the increase in stand-off distance, straight, wavy and continuously melted interface will be obtained in order [14]. The flyer plate is in an accelerated state before it touches the substrate, thus, when the stand-off distance is larger, the kinetic energy of the flyer plate will be transformed into heat energy near the interface [3,11,14]. Such sinusoidal shape of the interfaces can hinder the propagation of cracks and increase the strength of the welded dissimilar layers [20,21]. The upper-interface is more undulating than the lower-interface, and the measured average wavelength and amplitude of the upper-interface waves are 290 ± 20 μm and 75 ± 5 μm, respectively, those of the lower-interface are 231 ± 18 μm and 41 ± 3 μm. It may be due to the fact that the upper-interface has higher energy. In addition, there is a small amount of penetration at the upper interface, and it can be found from the energy dispersive X-ray detector (EDX) results (marked by black arrows in Figure 3b). This phenomenon suggests that the vortex is a melted zone. This could be attributed to the following reasons. Despite the fact that the copper has a very high value of heat conductivity, there was not enough time for the heat generated by the collision to be transferred to other position with lower temperatures at the moment of the explosion. The rapidly increasing temperature resulted in part of the copper and Q235 steel in the bonding interface to melt. The melting points of copper and steel are 1083 °C and 1500 °C, respectively. As a result, a number of the most severely deformed copper microstructures melted in the first place, and more liquid copper was mixed with the soften or even molten iron, which can be evidenced by the fact that copper occupies more volume fraction in the vortex region. Then, under the action of the shock wave, the molten copper and steel mixed together and flowed with the wave crest, and finally converge to the vortex region where the pressure in front of the wave is lowest [1,16,19]. Greenberg et al. [22,23] found the same phenomenon in aluminium–tantalum and copper–titanium explosive welding interfaces, vortex zones are formed near the cusps and valleys. Due to the low impact energy, we did not find a similar vortex zone at the lower-interface, as shown in Figure 3c.

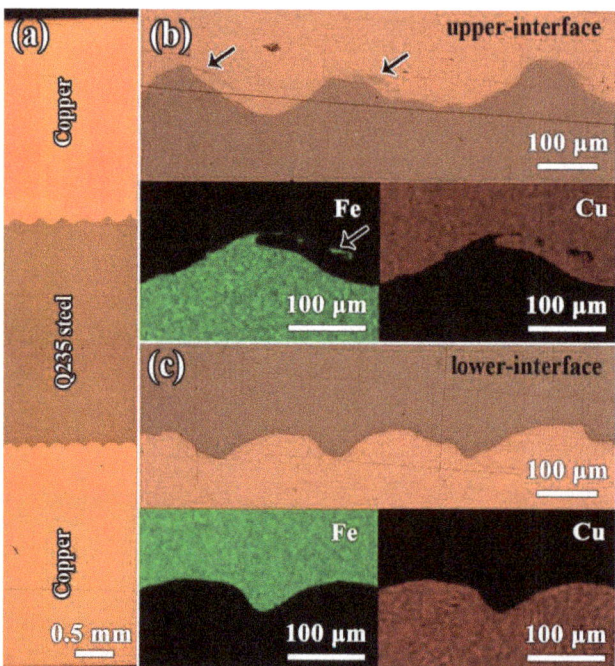

Figure 3. Optical microscope images (**a**) of the wave crest structure in copper/steel explosive bonding interface (**b,c**) are upper-interface and lower-interface, respectively. The insets in (**b,c**) are energy dispersive X-ray detector (EDX) maps and color codes.

For further understanding, the upper-interface and lower-interface are analyzed by EBSD, as shown in Figure 4. Figure 4a,b displays the microstructures of two interfaces. Inverse pole figure (IPF) indicates that the grains away from the interface in copper and steel are equiaxed, and the grain size is consistent with the metallographic statistics. However, the grains of steel have a certain selective orientation with <100> and <111> oriented parallel to the direction of the wave crest line. This could be the inheritance of the initial microstructure, considering the strain gradient distribution at the interface due to the explosion [12]. Two typical areas are selected and enlarged in Figure 4c,d, respectively. The upper-interface and lower-interface are composed of ultra-fine grains (UFGs) with thicknesses of ~80 μm and ~50 μm (the area marked by the yellow dotted line). The average grain size of UFG in domains of upper-interface and lower-interface are ~3.6 μm and ~4.8 μm. Different from conventional welding techniques, the copper/Q235 steel interface obtained via explosive welding without the appearance of coarse grains with poor-strength which are often the site of cracks sprouting and eventually severely degrade the mechanical properties of the welded interface. The formation of UFGs can be ascribed to the high temperature generated during the impact at the bonding interface and subsequent rapid cooling. The distribution of grain boundaries in Figure 4e,f suggests that the UFGs mainly contain high angle grain boundaries (HAGBs), while the density of low angle grain boundaries (LAGBs) is higher beyond the interface and decreases with increasing distance. Which is consistent with the decrease in hardness gradient along the weld interface reported in Liu et al. [12] and Zhou et al. [8]. Because high dislocation density and reduced grain size often lead to high hardness [24,25].

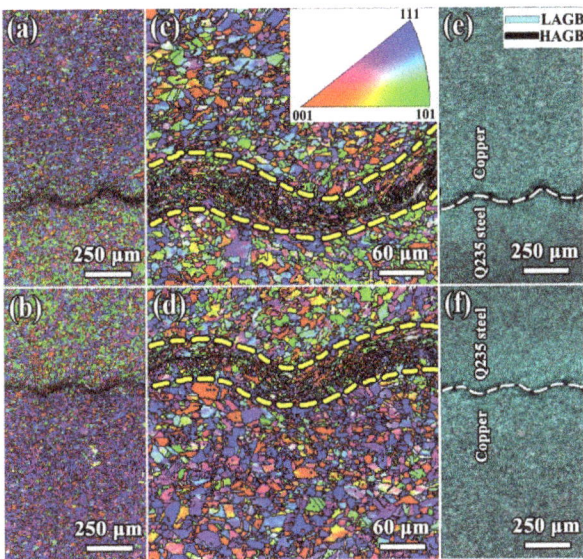

Figure 4. Inverse pole figures (**a**–**d**), grain boundaries (GBs) maps (**e**,**f**) of the wave crest structure in copper/Q235 steel explosive welding interface. (**a**,**c**,**e**) and (**b**,**d**,**f**) are upper-interface and lower-interface, respectively. The inset in (**c**) is color coded. For GB maps, black and cyan lines represent high angle grain boundaries (HAGBs) > 15° and low angle grain boundaries (LAGBs) between 2° and 15°, respectively.

To investigate the interfacial bonding strength of the copper/Q235 steel/copper explosive composite, the tensile-shear test was employed, and the load-displacement curves are shown in Figure 5. It can be found that the trend of the tensile-shear curve of the upper-interface is basically the same as that of the lower-interface. Shear strength (τ) is commonly used to measure the ability of a material to resist shear sliding, which is defined as $\tau = F_{max}/S$, where F_{max} and S are the maximum load and bonding surface area between the two notches [8]. The corresponding shear strengths of the upper-interface and lower-interfaces are higher than ~235 MPa and ~222 MPa, respectively. However, the energy consumed by the upper-interface and lower-interfaces is also almost equal during the tensile shear test because the area enclosed by the tensile-shear curve and the displacement axis is proportional to the energy consumed by the specimen during the deformation process [26,27].

Interestingly, the welded interfaces in the present work exhibited a different fracture behavior during tensile shear testing than in other literature [8,14]. Usually, failure of the welded interface occurs along the direction of the interface wave. This is due to the presence of defects such as cavities, cracks and brittle intermetallic at the interface zone. The discontinuities and stress concentration points can weaken the interface shear resistance [8]. However, during the experiment in this work, all fracture occurs on the copper side, the interface of the welded joint maintains the shape before the shear test (Figure 5). Fracture on the copper side indicates that the tensile-shear test is failed. In addition, the plastic deformation on the copper side resulted in much more slip traces and surface relief than the steel side which indicates that the plastic deformation during the tensile-shear test occurred mainly on the copper side. The reason for this phenomenon is that pure copper has a low strength. In other words, the shear strength of the upper-interface and lower-interface is much higher than 220 MPa. No obvious micro-cracks and fragmented brittle intermetallic compounds were found at the fracture (Figure 6a–d). From Figure 6c,d, it can be found that the plastic deformation on the copper side mainly occurs away from the interface, which is due to the fact that the UFGs can increase the coordination of plastic deformation in the local area and inhibit the generation of micro-cracks [26,27]. Thus, the explosive welding interface obtained in this work is excellent. This can also be judged from the fracture morphology of the fractured samples, where a large number of dimples indicate that ductile fracture occurred, and the size of the dimples is comparable to the grain size counted by EBSD, as shown in Figure 6e,f. Good quality of interfacial bonding will reduce the tendency of interfacial delamination in subsequent plastic processing [2].

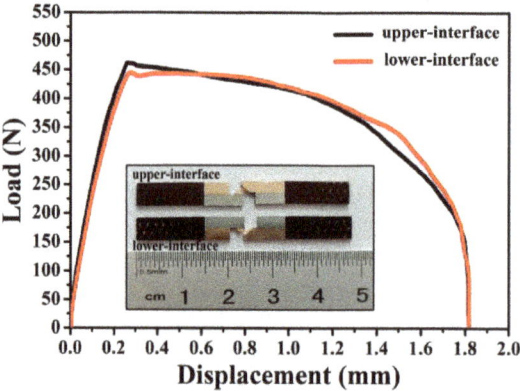

Figure 5. Comparison of the obtained load-displacement curves by tensile-shear tests for upper-face sample and lower-face sample, and the inset is the image of corresponding specimens after the tensile-shear test.

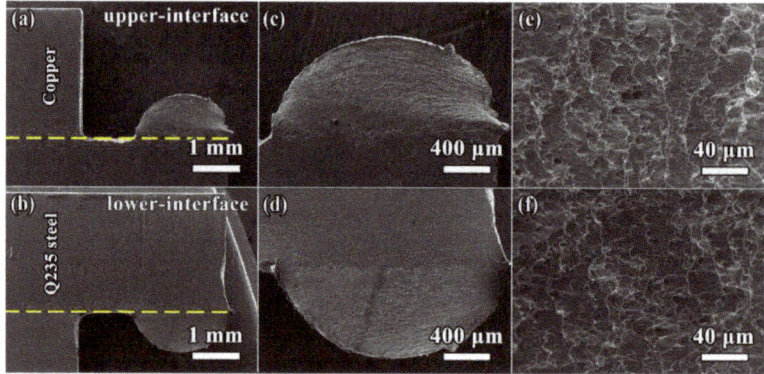

Figure 6. Scanning electron microscope (SEM) morphology of fractured tensile-shear specimens: (**a**–**d**) side view, (**e**,**f**) top view. The upper row and bottom row are upper-face and lower-face, respectively.

4. Conclusions

In this paper, the microstructures and properties of the interface for a copper/Q235 steel/copper composite plate prepared by explosive welding are studied. The interface between copper and Q235 steel presents a typical wavy structure with an amplitude ranging between 40 and 75 μm and a wavelength from 230 to 290 μm. The shear strength of the upper-interface and lower-interfaces of the welded copper/Q235 steel are higher than ~235 MPa and ~222 MPa, respectively. The specimens failed fully within the copper and not at the bonding interface, which is due to three factors: (1) no cavities and cracks at the interface; (2) the interface forms a metallurgical bonding including numerous UFGs, which can significantly improve the plastic deformation coordination at the interface and inhibit the generation of micro-cracks. Explosive welding can avoid the reduction of mechanical properties of welded joints due to defects such as coarse grains and cracks at the bonding interface. This research provides the experimental basis for sample preparation, microstructure observation and performance testing of the interface between two or more dissimilar metals. Furthermore, the simulation of interface microstructure and properties also has very important guiding significance for the bonding of two or more dissimilar metals. In addition, small-sized copper/Q235 steel/copper composite plates with excellent bonding interface were obtained in this study, and the preparation and deformation processes of large-sized composites suitable for engineering applications are yet to be investigated. They are the future directions of this work.

Author Contributions: J.L.: Investigation, writing–original draft, funding acquisition. Z.X.: investigation. Y.Z.: resources, funding acquisition. W.J.: investigation, resources. W.Q.: formal analysis, funding acquisition, writing–review & editing. Q.M.: formal analysis, software, writing–review & editing. Y.W.: investigation, funding acquisition. B.W.: formal analysis, funding acquisition. All authors have read and agreed to the published version of the manuscript.

Funding: This research was funded by National Natural Science Foundation of China grant number 52101030, 42202343, Natural Science Foundation of Anhui Province grant number 2208085QE125, the Scientific Research Foundation of Anhui Polytechnic University grant number HX202211007, Xjky2022024, Xjky2022032, the Open Research Found of Anhui Key Laboratory of High-performance Non-ferrous Metal Materials grant number YSJS-2023-01 and Major Research Development Program of Wuhu grant number 2022yf61.

Institutional Review Board Statement: Not applicable.

Informed Consent Statement: Not applicable.

Data Availability Statement: The data presented in this study are available on request from the corresponding author.

Conflicts of Interest: The authors declare that they have no known competing financial interest or personal relationship that could have appeared to influence the work reported in this paper.

References

1. Zhang, H.; Jiao, K.X.; Zhang, J.L.; Liu, J. Comparisons of the microstructures and micro-mechanical properties of copper/steel explosive-bonded wave interfaces. *Mater. Sci. Eng. A* **2019**, *756*, 430–441. [CrossRef]
2. Gladkovsky, S.V.; Kuteneva, S.V.; Sergeev, S.N. Microstructure and mechanical properties of sandwich copper/steel composites produced by explosive welding. *Mater. Charact.* **2019**, *154*, 294–303. [CrossRef]
3. Zhang, H.; Jiao, K.X.; Zhang, J.L.; Liu, J. Experimental and numerical investigations of interface characteristics of copper/steel composite prepared by explosive welding. *Mater. Des.* **2018**, *154*, 140–152. [CrossRef]
4. Wu, H.; Xiong, S.; Zhang, B.; Liu, Z.; Zhou, J.; Wang, C.; Xu, N. Influence of cold rolling parameters on bonding of copper/steel clad plate. *J. Plast. Eng.* **2021**, *28*, 75–82.
5. Sebastian, S.; Suyamburajan, V. Microstructural analysis of diffusion bonding on copper stainless steel. *Mater. Today Proc.* **2021**, *37*, 1706–1712. [CrossRef]
6. Danes, C.A.; Dumitriu, C.; Vizireanu, S.; Bita, B.; Nicola, I.M.; Dinescu, G.; Pirvu, C. Influence of carbon nanowalls interlayer on copper deposition. *Coatings* **2021**, *11*, 1395. [CrossRef]
7. Li, H.; He, Y.; Zhang, H.; Ma, T.; Li, Y. Study on the bonding mechanism of copper-low carbon steel for casting compounding process. *Metals* **2021**, *11*, 1818. [CrossRef]
8. Zhou, Q.; Liu, R.; Zhou, Q.; Chen, P.; Zhu, L. Microstructure characterization and tensile shear failure mechanism of the bonding interface of explosively welded titanium-steel composite. *Mater. Sci. Eng. A* **2021**, *820*, 141559. [CrossRef]
9. Sui, G.; Li, J.; Sun, F.; Ma, B.; Li, H. 3D finite element simulation of explosive welding of three-layer plates. *Sci. China Phys. Mech. Astron.* **2011**, *54*, 890–896. [CrossRef]
10. Yang, M.; Chen, D.; Zhou, H.; Xu, J.; Ma, H.; Shen, Z.; Zhang, B.; Tian, J. Experimental and numerical investigation of microstructure and evolution of TiNi Alloy/Q235 steel interfaces prepared by explosive welding. *J. Mater. Res. Technol.* **2021**, *15*, 5803–5813. [CrossRef]
11. Wang, J.; Lu, X.; Cheng, C.; Li, B.; Ma, Z. Improve the quality of 1060Al/Q235 explosive composite plate by friction stir processing. *J. Mater. Res. Technol.* **2020**, *9*, 42–51. [CrossRef]
12. Liu, L.; Jia, Y.F.; Xuan, F.Z. Gradient effect in the waved interfacial layer of 304L/533B bimetallic plates induced by explosive welding. *Mater. Sci. Eng. A* **2017**, *704*, 493–502. [CrossRef]

13. Abe, A. Numerical simulation of the plastic flow field near the bonding surface of explosive welding. *J. Mater. Process. Technol.* **1999**, *85*, 162–165. [CrossRef]
14. Guo, X.; Ma, Y.; Jin, K.; Wang, H.; Tao, J.; Fan, M. Effect of stand-off distance on the microstructure and mechanical properties of Ni/Al/Ni laminates prepared by explosive bonding. *J. Mater. Eng. Perform.* **2017**, *26*, 4235–4244. [CrossRef]
15. Ismail, N.; Jalar, A.; Bakar, M.A.; Safee, N.S.; Yusoff, W.Y.W.; Ismail, A. Microstructural evolution and micromechanical properties of SAC305/CNT/CU solder joint under blast wave condition. *Solder. Surf. Mount Technol.* **2020**, *33*, 47–56. [CrossRef]
16. Li, Z.; Xue, W.; Chen, Y.; Yu, W.; Xiao, K. Microstructure and grain boundary corrosion mechanism of pearlitic material. *J. Mater. Eng. Perform.* **2021**, *31*, 483–494. [CrossRef]
17. Zhang, F.; Mao, X.; Bao, S.; Zhao, G.; Zhao, S.; Deng, Z.; He, M.; Huang, F.; Qu, X. Microstructure evolution and its effects on the mechanical behavior of cold drawn pearlite steel wires for bridge cables. *J. Wuhan Univ. Technol. Mater. Sci. Ed.* **2022**, *37*, 96–103. [CrossRef]
18. Doherty, R.D.; Hughes, D.A.; Humphreys, F.J.; Jonas, J.J.; Jensen, D.J.; Kassner, M.E.; King, W.E.; Mcnelley, T.R.; Mcqueen, H.J.; Rollett, A.D. Current issues in recrystallization: A review. *Mater. Today* **1998**, *1*, 14–15. [CrossRef]
19. Pouraliakbar, H.; Khalaj, G.; Jandaghi, M.R.; Fadaei, A.; Ghareh-Shiran, M.K.; Shim, S.H.; Hong, S.I. Three-layered SS321/AA1050/AA5083 explosive welds: Effect of PWHT on the interface evolution and its mechanical strength. *Int. J. Press. Vessel. Pip.* **2020**, *188*, 104216. [CrossRef]
20. Durgutlu, A.; Gülenç, B.; Findik, F. Examination of copper/stainless steel joints formed by explosive welding. *Mater. Des.* **2005**, *26*, 497–507. [CrossRef]
21. Findik, F. Recent developments in explosive welding. *Mater. Des.* **2011**, *32*, 1081–1093. [CrossRef]
22. Greenberg, B.A.; Ivanov, M.A.; Kuzmin, S.V.; Lysak, V.I.; Besshaposhnikov, Y.P.; Pushkin, M.S.; Inozemtsev, A.V.; Patselov, A.M. Microstructures upon explosion welding and processes which prevent joining of materials. *Lett. Mater.* **2018**, *8*, 252–257. [CrossRef]
23. Grinberg, B.A.; Pushkin, M.S.; Patselov, A.M.; Inozemtsev, A.V.; Ivanov, M.A.; Slautin, O.V.; Besshaposhnikov, Y.P. The structure of molten zones in explosion welding (aluminium–tantalum, copper–titanium). *Weld. Inter.* **2017**, *31*, 384–393. [CrossRef]
24. Mao, Q.; Zhang, Y.; Liu, J.; Zhao, Y. Breaking material property trade-offs via macrodesign of microstructure. *Nano Lett.* **2021**, *21*, 3191–3197. [CrossRef] [PubMed]
25. Mao, Q.; Zhang, Y.; Guo, Y.; Zhao, Y. Enhanced electrical conductivity and mechanical properties in thermally stable fine-grained copper wire. *Commun. Mater.* **2021**, *2*, 46. [CrossRef]
26. Li, J.; Cao, Y.; Gao, B.; Li, Y.; Zhu, Y. Superior strength and ductility of 316L stainless steel with heterogeneous lamella structure. *J. Mater. Sci.* **2018**, *53*, 10442–10456. [CrossRef]
27. Li, J.; Fang, C.; Liu, Y.; Huang, Z.; Wang, S.; Mao, Q.; Li, Y. Deformation mechanisms of 304L stainless steel with heterogeneous lamella structure. *Mater. Sci. Eng. A* **2019**, *742*, 409–413. [CrossRef]

Disclaimer/Publisher's Note: The statements, opinions and data contained in all publications are solely those of the individual author(s) and contributor(s) and not of MDPI and/or the editor(s). MDPI and/or the editor(s) disclaim responsibility for any injury to people or property resulting from any ideas, methods, instructions or products referred to in the content.

Communication

Investigation of the Microstructures and Properties of B-Bearing High-Speed Alloy Steel

Jingqiang Zhang [1,*], Penghui Yang [2,*] and Rong Wang [2]

1. School of Mechanical and Energy Engineering, Zhejiang University of Science and Technology, Hangzhou 310023, China
2. State Key Lab Adv Proc & Recycling Nonferrous Met, Lanzhou University of Technology, Lanzhou 730050, China
* Correspondence: jingqiangz@126.com (J.Z.); yangph@lut.edu.cn (P.Y.)

Abstract: This work aims to research the influence of boron and quenching temperature on the microstructures and performances of boron-bearing high-speed alloy steel. The results showed that the hardness and wear resistance of boron-bearing high-speed alloy steel were improved after increasing the boron content. The volume fraction of boron-rich carbide gradually decreased, and the hardness increased significantly with the rise in quenching temperature. The highest comprehensive mechanical properties were obtained for samples quenched at 1040 °C. The TEM results showed the boron-rich carbide was $M_7(C, B)_3$ with an HCP structure, and the precipitated particles were $M_{23}(C, B)_6$ with an FCC structure after tempering. This work may help improve the wear resistance of materials in the field of surface coatings.

Keywords: boron-bearing high-speed alloy steel; properties; quenching temperature; boron-rich carbide

Citation: Zhang, J.; Yang, P.; Wang, R. Investigation of the Microstructures and Properties of B-Bearing High-Speed Alloy Steel. *Coatings* 2022, 12, 1650. https://doi.org/10.3390/coatings12111650

Academic Editors: Michał Kulka and Ludmila B. Boinovich

Received: 19 September 2022
Accepted: 28 October 2022
Published: 31 October 2022

Publisher's Note: MDPI stays neutral with regard to jurisdictional claims in published maps and institutional affiliations.

Copyright: © 2022 by the authors. Licensee MDPI, Basel, Switzerland. This article is an open access article distributed under the terms and conditions of the Creative Commons Attribution (CC BY) license (https://creativecommons.org/licenses/by/4.0/).

1. Introduction

The roller is an important part of the production of hot rolling or cold rolling [1,2]. The working environment of the roller is very harsh, which requires good high-temperature oxidation resistance, high impact toughness, and excellent wear resistance [3,4]. High-speed alloy steel is often used in the roller due to its high hardness, good impact toughness, and excellent wear resistance [5,6]. However, a lot of alloying elements are also added to the high-speed alloy steel to enhance the wear resistance, which has increased the production cost, resulting in a huge waste of metallic resources. Surface coating technology is also often used to enhance the properties of a roller, but the wear resistance and the hardness of the substrate are also critical. Therefore, this work sets the scene for the remaining laser cladding works by studying the properties of the substrate.

To reduce the consumption of energy, low-cost materials are important to use in the equipment, particularly in wear-resistant materials. The element boron is often added to steel to replace expensive elements and improve wear resistance, as boron can combine with C, Fe, Cr, Mo, and other elements to form hard phases with good stability and high hardness. Moreover, the addition of boron can allow for its distribution into the matrix to enhance the hardenability of materials and, best of all, reduce the cost compared to other elements [7–9].

Chen et al. [8] researched the microstructures and properties of white cast iron with high boron content and found that M_2B-type eutectic compounds formed after adding boron. XRD, EDX, and TEM results showed that the $M'_{0.9}Cr_{1.1}B_{0.9}$ eutectic compounds appeared in the matrix of white cast irons when adding about 4 wt.% chromium. Fu et al. [9] investigated the microstructures, performance, and wear resistance of boron-bearing high-speed steel. They found that the casting-state structure of boron-bearing high-speed steel consisted of $M_2(B,C)$ carbides, $M_{23}(B,C)_6$, α-Fe, $M_3(B_{0.7}C_{0.3})$, TiC, and a small quantity of retained austenite. The impact toughness and hardness reached 80–85 kJ/cm^2 and 65–67

HRC, respectively. However, this high-speed steel with low carbon and high alloy content did not achieve high wear resistance, and the W and Ti elements were expensive.

Therefore, in this work, boron will be added to high-speed alloy steel with high carbon content to study the influence of boron and heat treatment processes on high-speed alloy steel. More importantly, the developed high-speed alloy steel will be used to investigate if the use of expensive elements could be decreased by adding boron to reduce costs.

2. Experiment Details

2.1. Material Preparation

The boron-bearing high-speed alloy steel was prepared in a 40 kg medium-frequency induction furnace. All raw materials were melted, and ferroboron was added into the melt as the temperature reached 1540 °C. The melt was poured into sand molds when the temperature rose to 1610 °C, and then Y-block ingots were obtained according to the standard of ASTM A781/A 781-M95. Table 1 shows the chemical constituent of the testing alloy steel.

Table 1. The chemical constituent of the testing alloy steel (wt.%).

Alloy	C	B	Mo	Si	Mn	Cr	V	S	P	Fe
A	1.19	1.03	1.36	0.87	1.13	3.50	1.13	0.011	0.024	Bal.
B	1.23	1.59	1.41	0.92	1.24	3.48	1.05	0.016	0.025	Bal.

The heat treatment processes of boron-bearing high-speed alloy steel were as follows: the alloy steel was austenitized at different temperatures for 1 h, then quenched by air. After that, the samples were tempered at 500 °C for 3 h and then cooled by air.

2.2. Microstructural Investigation

The microstructures of the samples were investigated by a scanning electron microscope (SEM, JSM-6510, JEOL, Tokyo, Japan) (secondary electron detector), an optical microscope (OM, Olympus BX51, Olympus, Tokyo, Japan), an energy dispersive spectroscope (EDS, Oxford Instruments, Oxford, London, England) with an acceleration voltage of 15 kV, a transmission electron microscope (TEM, JEM-2100F, JEOL, Tokyo, Japan) [10] with an energy-dispersive X-ray (EDX, Oxford Instruments, 80 mm^2 X-max SDD, Oxford, London, England), and an electron probe X-ray micro-analyzer (EPMA, JXA-8230, EOL, Tokyo, Japan). The samples were polished and then etched by a 5 g $CuSO_4$ + 20 mL HCl + 20 mL H_2O solution for metallographic observation. The volume fraction of the boron-rich carbide (B-rich carbide) was measured by Image-Pro software (1.0, Tokyo, Japan). Fifty different fields of view were selected and calculated the average value was the final result (the standard deviation was less than 4). The phase composition was analyzed by X-ray diffraction (XRD, Rigaku D/Max-2400× diffractometer, Rigaku, Tokyo, Japan) with a 2θ range of 10°–90° and a scanning speed of 20°/min.

2.3. Properties Investigation

The hardness of boron-bearing high-speed alloy steel was measured by the HR-150A Rockwell hardness tester (TIME, Beijing, China), and the average value was calculated as the resulting value (the standard deviation was less than 5). The microhardness measurements of boron-bearing high-speed alloy steel samples were carried out by a MICRO MET-5103 Vickers hardness tester with a load of 1.96 N. According to the standard of ASTM E384-1999, the unnotched Charpy impact specimens were measured by a JBW-300 capacity impact testing machine with an impact energy of 150 J at room temperature. The size of the samples for the impact test was 10 mm × 10 mm × 55 mm. Wear loss was measured by an MM-200 block-on-ring wear testing machine with a wheel rotation speed of 300 r/min, a load of 300 N, and a test duration of 60 min. The size of the specimen for the wear test was 10 mm × 10 mm × 10 mm. The grinding ring with a material of GCr15 steel, a hardness of 60 HRC, and a size of φ40 mm × 10 mm (thickness) was selected to rub the specimen for the wear test. In order to compare wear resistance, the wear loss was measured by the electronic balance (TG328B), and then the average value of three samples was calculated as the final weight loss result. The worn surfaces of the specimens were analyzed by a VK-9710 color 3D laser scanning microscope (Keyence, Osaka, Japan) and SEM.

3. Results and Discussion

3.1. As-Cast Microstructures

Figure 1 shows the as-cast microstructures of the boron-bearing high-speed alloy steel (B-bearing high-speed alloy steel). The solidification microstructures of the two boron-bearing high-speed alloy steel are mainly constituted of the matrix and the coarse net-like B-rich carbides. Thereinto, the matrix contains ferrite mainly, some pearlites, and less martensite. The area of B-rich carbide was calculated by Image-Pro software (Version 6.0), and the results showed that the volume fraction of B-rich carbide in alloy A and alloy B was 23.1 and 28.9 vol.%, respectively. Evidently, the quantity of B-rich carbide in alloy B was more than in alloy A, indicating that the B-rich carbides increased with the increase in boron content.

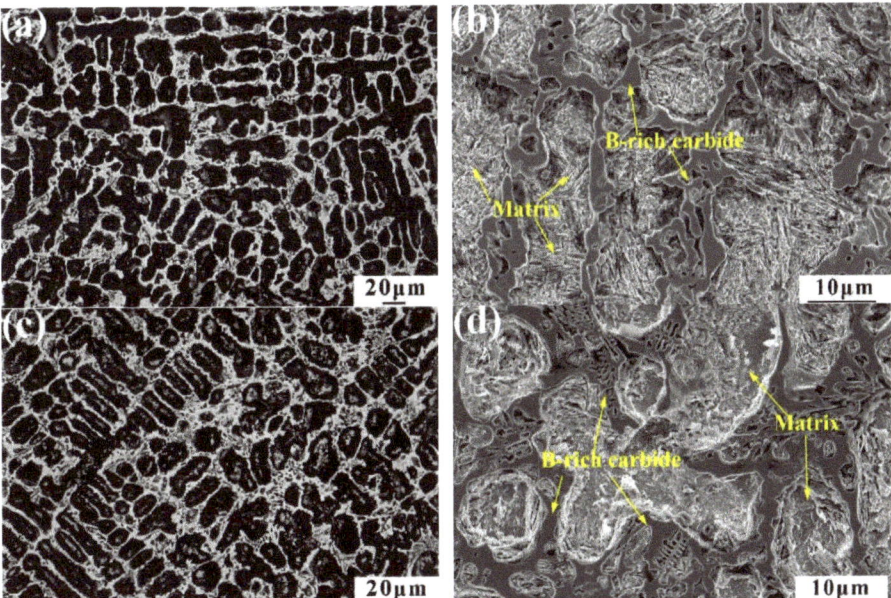

Figure 1. Microstructures of alloy steel as-cast: (**a**) OM result of alloy A; (**b**) SEM result of alloy A; (**c**) OM result of alloy B; (**d**) SEM result of alloy B.

In order to study the element constitution and the distribution of B-rich carbide in the two alloys, the surface of specimens was observed by an SEM with an EDS. Figures 2 and 3 show the surface scanning results of alloys A and B, respectively. Table 2 shows EPMA test results of B-rich carbides. The results showed that the elements contained in the carbides are C, B, Cr, Mo, etc., indicating that the Cr and Mo elements were dissolved in the B-rich carbide. The microhardness of the B-rich carbide in alloy A and alloy B is 644 and 731 $HV_{0.2}$, which indicates the hardness of B-rich carbide increased with B content. In addition, the microhardness of the matrix in alloy A and alloy B is 255 and 246 $HV_{0.2}$, indicating that the matrix hardness did not significantly change after increasing the B content.

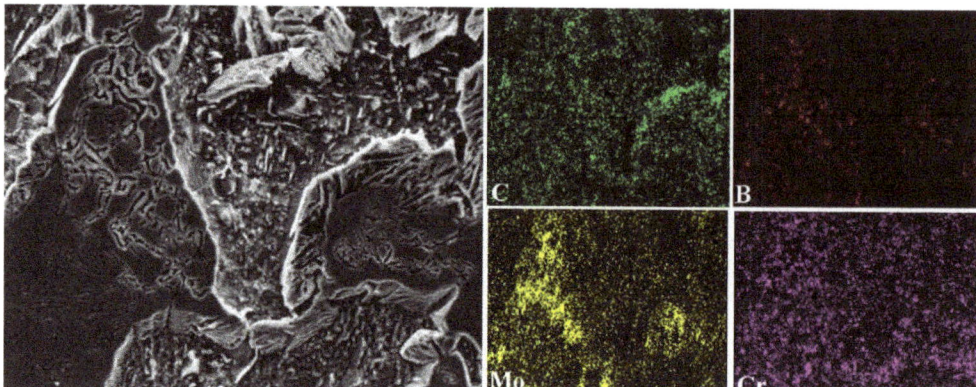

Figure 2. EDS analysis of as-cast alloy A.

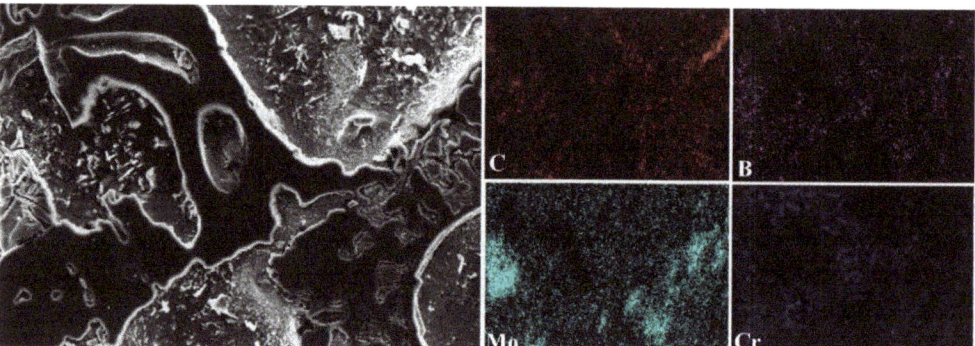

Figure 3. EDS analysis of as-cast alloy B.

Table 2. EPMA test results of B-rich carbides (at.%).

Alloy	C	B	Mo	Si	Mn	Cr	V	Fe
A	13.45	18.33	2.88	0.96	0.24	6.77	1.79	55.58
B	11.26	22.69	3.15	0.63	0.51	5.24	1.81	54.71

3.2. Effect of Quenching Temperature on Microstructures and Properties

Figure 4 shows the microstructure of alloy A after quenching at different temperatures. After quenched at 960 and 1000 °C, the matrix of the boron-bearing high-speed alloy steel contained ferrite and less pearlite, as shown in Figure 4a–d. The matrix changed from ferrite to martensite when the quenching temperature exceeded 1040 °C (see Figure 4e,f). With the increase in quenching temperature, the matrix did not change, as shown in Figure 4a–d, indicating that the high quenching temperature could obtain the martensitic matrix.

Figure 5 shows the microstructure of alloy B after quenching at different temperatures. Similar to the case of alloy A, the matrix of boron-bearing high-speed alloy steel contained ferrite and less pearlite after quenching at relatively low quenching temperatures, as shown in Figure 5a–d. After quenching at relatively high quenching temperatures (more than 1040 °C), the matrix of the boron-bearing high-speed alloy steel was mainly composed of martensite.

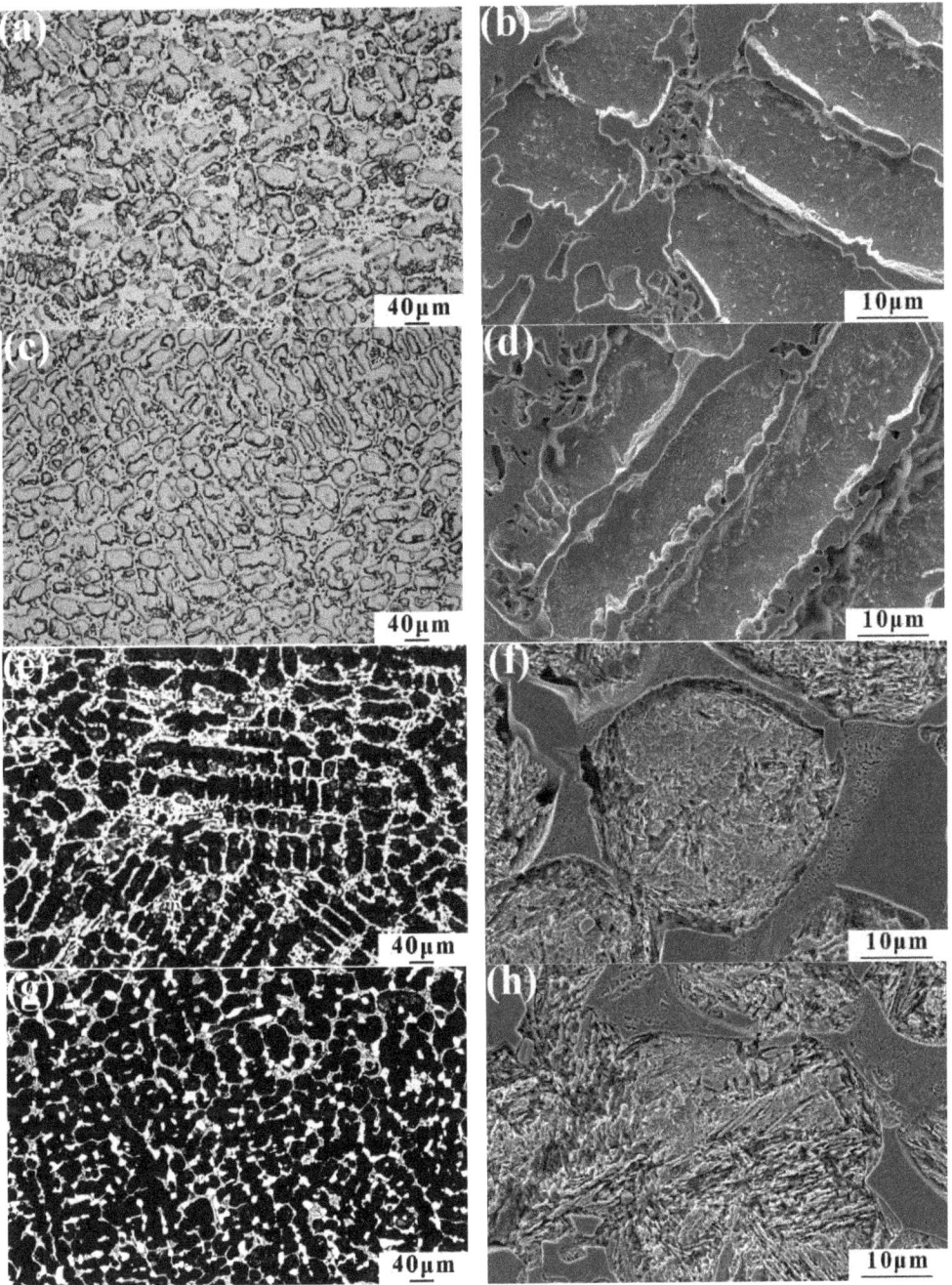

Figure 4. Microstructures of alloy A quenched at different temperature (°C): (**a**,**b**) 960; (**c**,**d**) 1000; (**e**,**f**) 1040; (**g**,**h**) 1080.

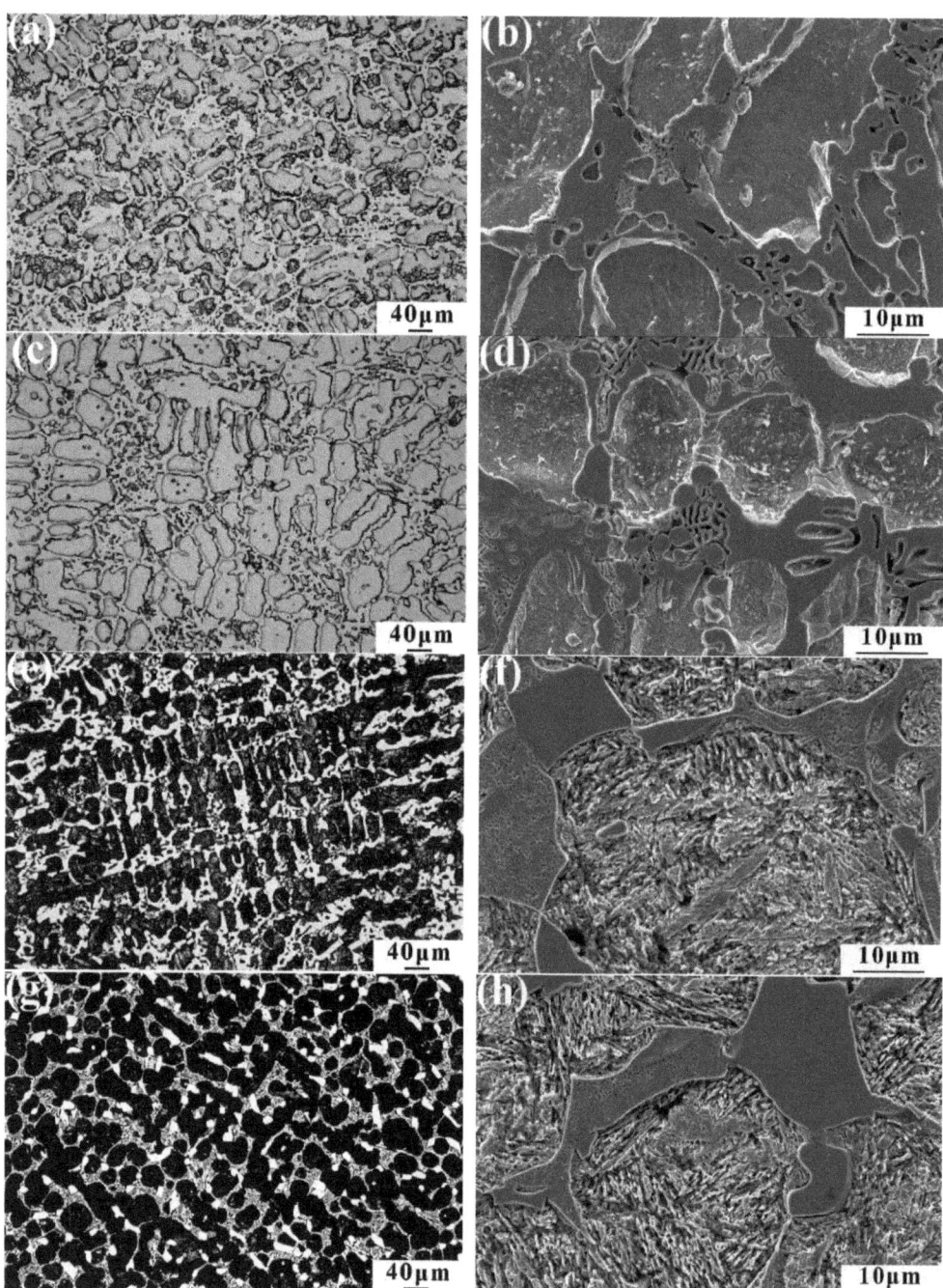

Figure 5. Microstructures of alloy B after quenching at different temperature (°C): (**a**,**b**) 960; (**c**,**d**) 1000; (**e**,**f**) 1040; (**g**,**h**) 1080.

To further analyze the influence of quenching temperature on B-rich carbide, the volume fraction of B-rich carbide in B-bearing high-speed alloy steel quenched at different temperatures was investigated. The computed results are shown in Figure 6. As the quenching temperature increased, the volume fraction of B-rich carbide gradually decreased. Alloy A and alloy B showed the same trend. High temperatures promoted the increase of boron solid solubility in austenite, which caused the elements in the B-rich carbide to diffuse into the matrix, resulting in the dissolution of the B-rich carbide. It is also worth noting that the volume fraction of B-rich carbide in alloy A was always smaller than in alloy B.

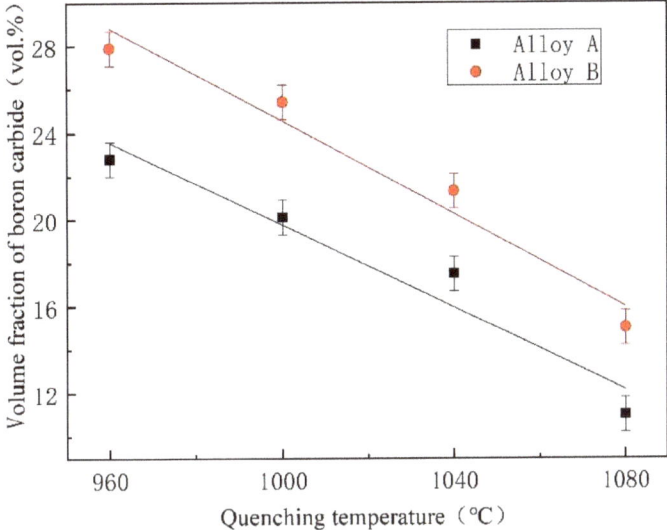

Figure 6. The volume fraction of B-rich carbide in boron-bearing high-speed alloy steel quenched at different temperatures.

Figure 7 shows the influence of quenching temperature on microhardness in the alloy steel matrix. The microhardness of the matrix of alloy A and alloy B increased significantly with the increase in quenching temperature. When quenched at 1080 °C, the microhardness of alloy A and alloy B reached the highest values, 720 and 780 $HV_{0.2}$, respectively. When the quenching temperature was relatively low, the matrix of the boron-bearing high-speed alloy steel was mainly composed of ferrite. After the quenching temperature reached 1080 °C, the ferrite matrix became martensite, which improved the hardness of the alloy. Furthermore, the elements in B-rich carbide dissolved into the matrix, which led to an increase in matrix hardness.

Figure 8 shows the properties of the boron-bearing high-speed alloy steel after quenching at different temperatures. The macro-hardness trend was similar to that of microhardness. Although the increase in the quenching temperature led to a decrease in B-rich carbide, the increase in matrix hardness promoted an increase in microhardness. The impact toughness gradually increased with the rise in quenching temperature. The structure of the boron-bearing high-speed alloy steel contained ferrite, less pearlite, and a large volume fraction of carbides at the low quenching temperature, so the hardness and impact toughness were lower. The appearance of martensite promoted an increase in hardness, and a decrease in carbides promoted an increase in impact toughness. However, the appearance of a large amount of martensite could destroy the impact toughness. For the two alloys, the quenching temperature to obtain the optimal comprehensive mechanical properties was 1040 °C. Therefore, alloys A and B were treated with a quenching temperature of 1040 °C to study the microstructure and wear resistance after tempering.

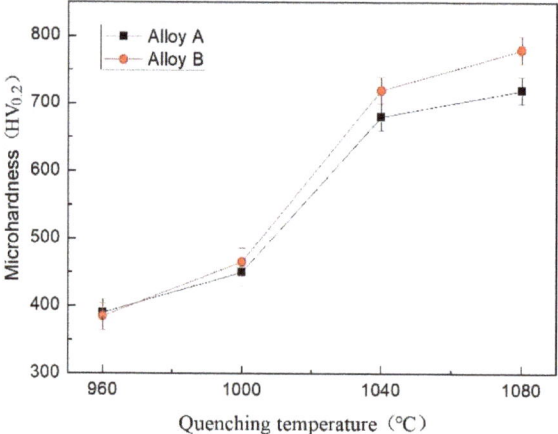

Figure 7. The microhardness in matrix of alloy steel.

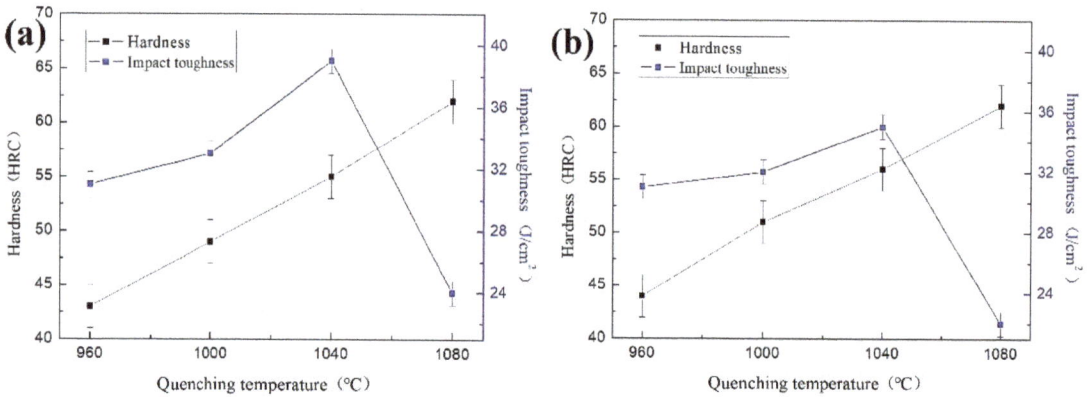

Figure 8. Hardness and impact toughness of alloy steel: (**a**) alloy A; (**b**) alloy B.

3.3. Microstructure and Wear Resistance after Tempering

Figure 9 shows the microstructures of alloys A and B in the heat-treated condition. The microstructure of the B-bearing high-speed alloy steel comprises precipitated particles, martensite matrix, and B-rich carbide after tempering at 500 °C for 3 h. It is worth noting that the number of particles was relatively small. The total volume fraction of B-rich carbide and precipitated particles were investigated by Image-Pro software, and the total volume fraction of alloy A and B were 19.2 and 23.3 vol.%, respectively, indicating that the total volume fraction of alloy A was slightly less than alloy B.

In order to further analyze the types of B-rich carbides and precipitated particles, the microstructures of alloy B were investigated by a TEM with an EDX. The results of the TEM micrographs and elemental analysis are shown in Figure 10 and Table 3, respectively. The B-rich carbide is $M_7(C, B)_3$ with a hexagonal close-packed (HCP) structure [11], and the lattice constants are a = 1.38 nm and b = 0.43 nm, as shown in Figure 10a. The precipitated particles are $M_{23}(C, B)_6$ with a face-centered cubic (FCC) structure [11], and the lattice constant is a = 1.07 nm, as shown in Figure 10b. Figure 11 shows the XRD results of the boron-bearing high-speed alloy steel after tempering. The matrix contained two kinds of carbides: $M_{23}(C, B)_6$ and $M_7(C, B)_3$, which is in agreement with the TEM results. In addition, the results of the high-resolution transmission electron microscopy (HR-TEM) in Figure 10c also provided support.

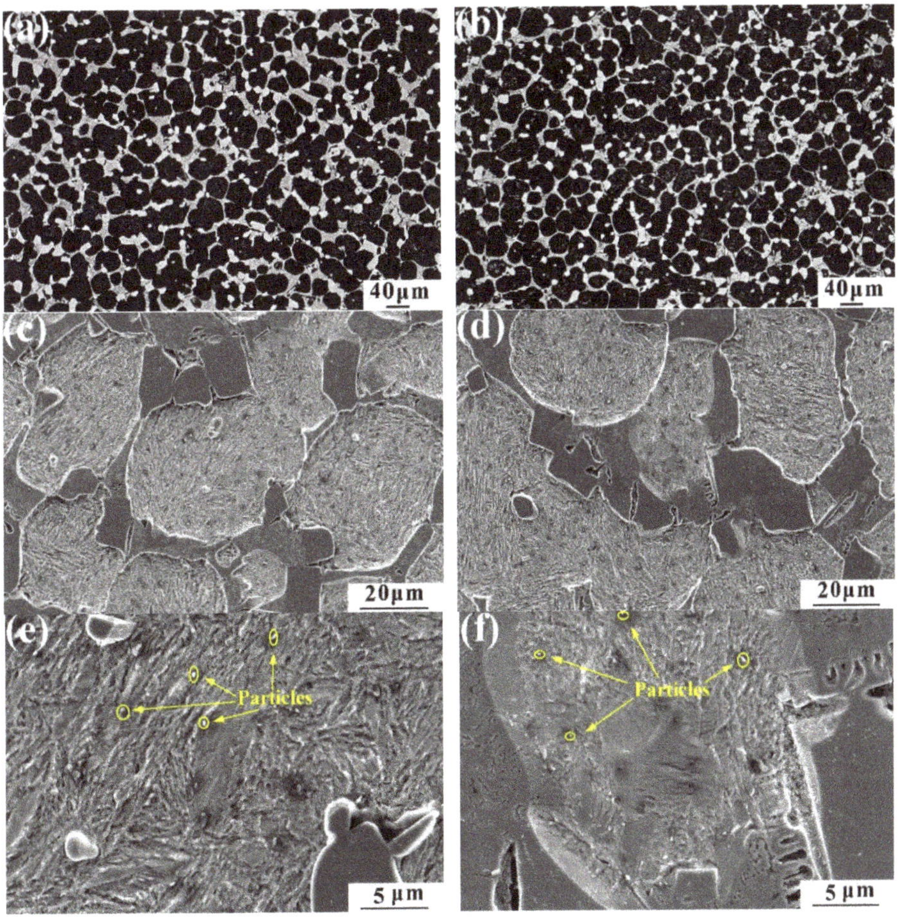

Figure 9. Microstructure of boron-bearing high-speed alloy steel after tempering: OM results (**a**), SEM results (**c**) and SEM results at high magnification (**e**) of alloy A; OM results (**b**), SEM results (**d**) and SEM results at high magnification (**f**) of alloy B.

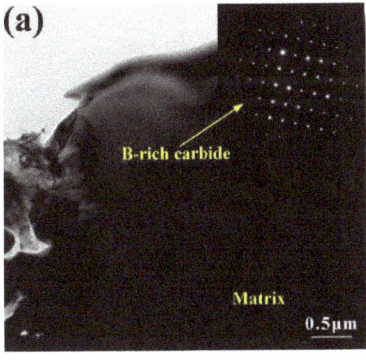
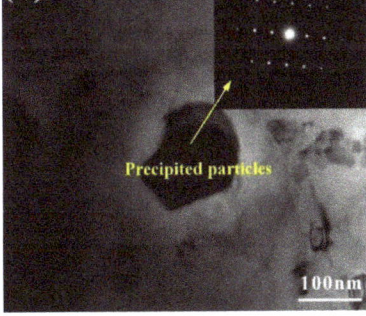

Figure 10. *Cont.*

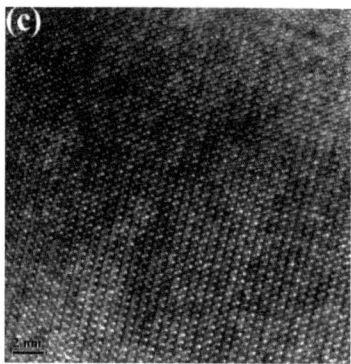

Figure 10. Bright-field TEM micrographs and corresponding selected area diffraction patterns (SADPs) of B-rich carbide and precipitated particles: (**a**) bright-field TEM micrographs and SADPs of B-rich carbide; (**b**) bright-field TEM micrographs and SADPs of precipitated particles; (**c**) HR-TEM image of precipitated particles.

Table 3. EDX test results of TEM samples (at%).

Phase	C	B	Mo	Si	Mn	Cr	V	Fe
B-rich carbides	11.68	19.65	3.56	1.12	0.56	5.69	1.52	56.13
Precipited particles	9.88	13.21	2.64	1.35	0.94	6.99	2.56	62.43

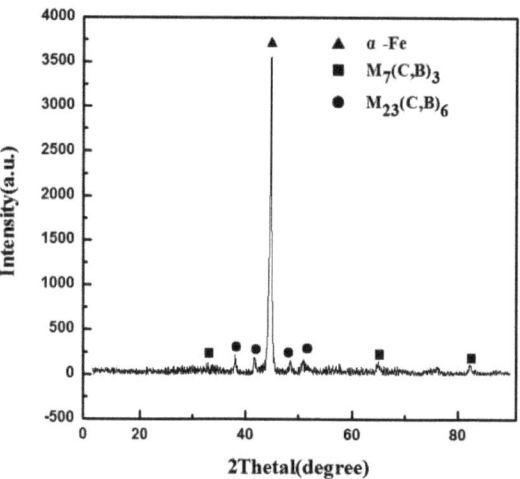

Figure 11. XRD result of B-bearing high-speed alloy steel after tempering.

The wear resistance of alloys A and B was researched by a wear tester, and the reciprocal of the wear loss was expressed as the wear resistance. The wear loss of alloys A and B are shown in Figure 12. The results show the wear resistance of alloy B exceeded alloy A, which indicates that the wear resistance of boron-bearing high-speed alloy steel increased after increasing the B content. The increase in wear resistance is mainly attributed to the increase in the volume fraction of carbides.

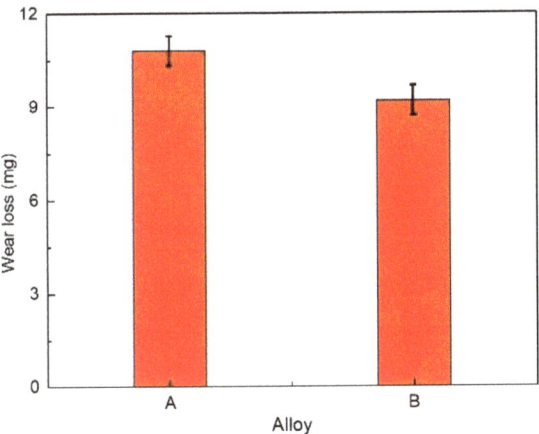

Figure 12. Wear loss of alloy A and alloy B.

The worn surface of the alloy steel was observed by an SEM, as shown in Figure 13a,b. The spalling and groove could be observed on the worn surface of the boron-bearing high-speed alloy steel, indicating that the wear mechanism of boron-bearing high-speed alloy steel may be abrasive wear. To reveal the wear mechanism of boron-bearing high-speed alloy steel, the cross-section of alloys A and B after wear was observed by an SEM, see Figure 13c,d. The B-rich carbide could protect the matrix, indicating that the wear resistance is significantly enhanced after increasing the B content.

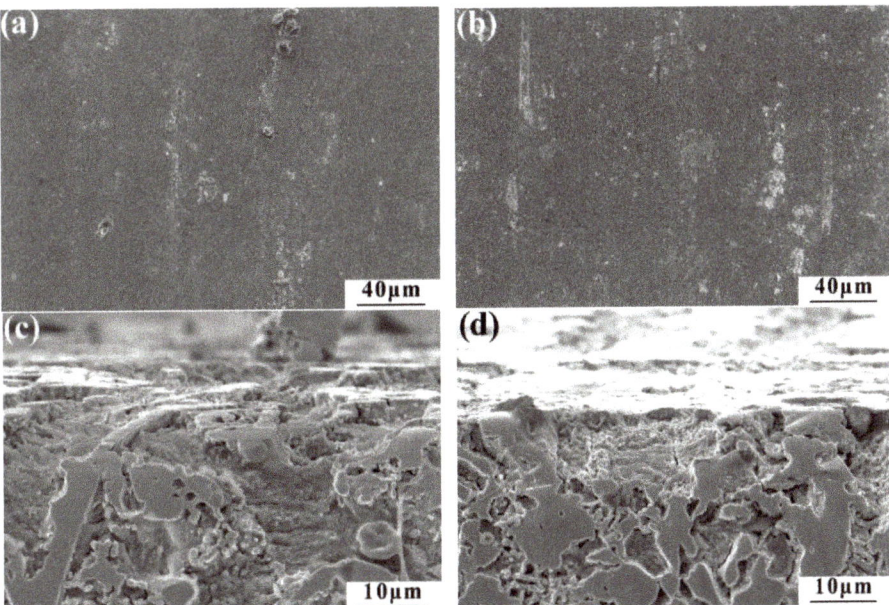

Figure 13. SEM images of B-bearing high-speed alloy steel worn surfaces: (**a**) SEM images of alloy A worn surfaces; (**b**) SEM images of alloy B worn surfaces; (**c**) SEM image of cross-section of alloy A after wear; (**d**) SEM image of cross-section of alloy B after wear.

To analyze the wear resistance of B-bearing high-speed alloy steel, the worn surface was observed by a 3D laser scanning microscope, see Figure 14. Due to the existence of B-rich carbide, the shedding was easily formed during the process of wear. Additionally, shallow grooves and low roughness occurred on the surface of alloy B because the wear resistance of alloy B is better than alloy A.

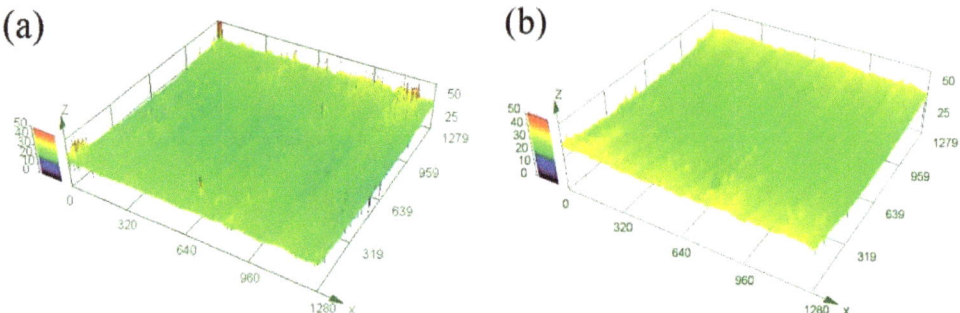

Figure 14. 3D laser morphologies of B-bearing high-speed alloy steel: (**a**) alloy A; (**b**) alloy B.

4. Conclusions

In this paper, boron was added to high-speed alloy steel with high carbon content to study the effects of boron and heat treatment on high-speed alloy steel. The conclusions obtained are as follows:

(1) With the increase in austenitizing temperature, the quantity of B-rich carbide gradually decreased. The comprehensive mechanical properties were optimal after the alloy steel was austenitized at 1040 °C for 1 h, quenched in air, and tempered at 500 °C for 3 h.
(2) TEM results showed the B-rich carbide is $M_7(C, B)_3$ with an HCP structure, and the precipitated particles are $M_{23}(C, B)_6$ with an FCC structure after tempering.
(3) The wear resistance of boron-bearing high-speed alloy steel was enhanced by increased B content.

Author Contributions: Conceptualization, J.Z. and P.Y.; methodology, J.Z.; software, R.W.; investigation, R.W. and P.Y. All authors have read and agreed to the published version of the manuscript.

Funding: This research received no external funding.

Institutional Review Board Statement: Not applicable.

Informed Consent Statement: Not applicable.

Data Availability Statement: The datasets generated during and/or analyzed during the current study are available from the corresponding author on reasonable request.

Conflicts of Interest: The authors declare no conflict of interest.

References

1. Gonçalves, J.L., Jr.; de Mello, J.D.B.; Costa, H.L. Wear in cold rolling milling rolls: A methodological approach. *Wear* **2019**, *426*, 1523–1535. [CrossRef]
2. Ghazanlou, S.I.; Eghbali, B.; Petrov, R. Study on the microstructural and texture evolution of Hot Rolled Al7075/graphene/carbon nanotubes reinforced composites. *Mater. Chem. Phys.* **2021**, *257*, 123766. [CrossRef]
3. Zhang, B.; Yang, C.; Sun, Y.; Li, X.; Liu, F. The microstructure, mechanical properties and tensile deformation mechanism of rolled AlN/AZ91 composite sheets. *Mater. Sci. Eng. A* **2019**, *763*, 138118. [CrossRef]
4. Aljabri, A.; Jiang, Z.Y.; Wei, D.B.; Wang, X.D.; Tibar, H. Thin strip profile control capability of roll crossing and shifting in cold rolling mill. *Mater. Sci. Forum* **2013**, *773*, 70–78. [CrossRef]
5. Pellizzari, M.; Molinari, A.; Straffelini, G. Tribological behaviour of hot rolling rolls. *Wear* **2005**, *259*, 1281–1289. [CrossRef]

6. Jiangtao, D.; Zhiqiang, J.; Hanguang, F. Effect of RE-Mg complex modifier on structure and performance of high speed steel roll. *J. Rare. Earths.* **2007**, *25*, 259–263. [CrossRef]
7. Yi, Y.; Xing, J.; Wan, M.; Yu, L.; Lu, Y.; Jian, Y. Effect of Cu on microstructure, crystallography and mechanical properties in Fe-B-C-Cu alloys. *Mater. Sci. Eng. A* **2017**, *708*, 274–284. [CrossRef]
8. Chen, X.; Li, Y.; Zhang, H. Microstructure and mechanical properties of high boron white cast iron with about 4 wt% chromium. *J. Mater. Sci.* **2010**, *46*, 957–963. [CrossRef]
9. Fu, H.; Ma, S.; Hou, J.; Lei, Y.; Xing, J. Microstructure and properties of cast Boron-bearing high speed steel. *J. Mater. Eng. Perform.* **2013**, *22*, 1194–1200. [CrossRef]
10. Wang, K.; Du, D.; Liu, G.; Pu, Z.; Chang, B.; Ju, J. High-temperature oxidation behaviour of high chromium superalloys additively manufactured by conventional or extreme high-speed laser metal deposition. *Corros. Sci.* **2020**, *176*, 108922. [CrossRef]
11. Ma, S.; Xing, J.; Fu, H.; Gao, Y.; Zhang, J. Microstructure and crystallography of borides and secondary precipitation in 18 wt.% Cr–4 wt.% Ni–1 wt.% Mo–3.5 wt.% B–0.27 wt.% C steel. *Acta Mater.* **2012**, *60*, 831–843. [CrossRef]

Article

Microstructure and Corrosion Resistance of Two-Dimensional TiO$_2$/MoS$_2$ Hydrophobic Coating on AZ31B Magnesium Alloy

Longjie Lai [1,2], Heng Wu [1,3], Guobing Mao [1,2], Zhengdao Li [4,*], Li Zhang [5] and Qi Liu [1,2,*]

1. School of Materials Science and Engineering, Anhui Polytechnic University, Wuhu 241000, China
2. Anhui Key Laboratory of High-Performance Non-ferrous Metal Materials, Anhui Polytechnic University, Wuhu 241000, China
3. School of Mechanical Engineering, Anhui Institute of Information Technology, Wuhu 241100, China
4. Chemistry and Pharmaceutical Engineering College, Nanyang Normal University, Nanyang 473061, China
5. Faculty of Institute of Photoelectronics Thin Film Devices and Technique of Nankai University, Nankai University, Tianjin 300071, China
* Correspondence: nylzd@nynu.edu.cn (Z.L.); modieer_67@ahpu.edu.cn (Q.L.)

Abstract: The corrosion resistance of magnesium alloys can be effectively improved by surface treatment. In this study, a hydrophobic two-dimensional (2D) TiO$_2$/MoS$_2$ nanocomposite coating was fabricated on AZ31B magnesium alloy by an electrophoretic deposition method. The corrosion resistance of the coating was evaluated using potentiodynamic polarization and electrochemical impedance spectroscopy analyses. After being modified by a silane coupling agent (KH570), the TiO$_2$/MoS$_2$ coating changed from hydrophilic to hydrophobic, and the static water contact angle increased to 131.53°. The corrosion experiment results indicated that the hydrophobic 2D TiO$_2$/MoS$_2$ coating had excellent anticorrosion performance (corrosion potential: E$_{corr}$ = −0.85 V$_{Ag/AgCl}$, and corrosion current density: I$_{corr}$ = 6.73 × 10^{-8} A·cm^{-2}). TiO$_2$/MoS$_2$ films have promising applications in magnesium alloy corrosion protection.

Keywords: magnesium alloys; electrophoretic deposition; TiO$_2$/MoS$_2$; corrosion resistance

Citation: Lai, L.; Wu, H.; Mao, G.; Li, Z.; Zhang, L.; Liu, Q. Microstructure and Corrosion Resistance of Two-Dimensional TiO$_2$/MoS$_2$ Hydrophobic Coating on AZ31B Magnesium Alloy. *Coatings* 2022, 12, 1488. https://doi.org/10.3390/coatings12101488

Academic Editor: Charafeddine Jama

Received: 23 August 2022
Accepted: 1 October 2022
Published: 6 October 2022

Publisher's Note: MDPI stays neutral with regard to jurisdictional claims in published maps and institutional affiliations.

Copyright: © 2022 by the authors. Licensee MDPI, Basel, Switzerland. This article is an open access article distributed under the terms and conditions of the Creative Commons Attribution (CC BY) license (https://creativecommons.org/licenses/by/4.0/).

1. Introduction

Magnesium alloys, widely used in the aerospace and automotive industries, have the advantages of high specific strength, low density, good manufacturability and recyclability, and abundant resources [1–4]. However, due to the relatively active chemical properties of magnesium, low standard electrode potential, and poor protection ability of natural oxide film, the corrosion resistance of magnesium alloys is poor [5–7], which limits their wider application. Therefore, improving the corrosion resistance of magnesium alloys can greatly expand their practical application. The commonly used methods to improve the corrosion resistance of magnesium alloys include chemical conversion coating [8], plasma electrolytic oxidation [9], laser surface melting [10], shot peening [11], cold spraying [12], etc. These methods can improve the corrosion resistance of magnesium alloys, but the process is complex, costly, energy-intensive, and poses various environmental issues. On the other hand, constructing a thin film on the surface of magnesium alloys by electrodeposition can effectively slow down the corrosion rate of materials and has the advantages of low cost, convenient operation, rapid film formation, and ease of control [13,14].

Two-dimensional materials have attracted more attention for their potential applications in optics, electrochemical energy storage, biosensing, and other fields due to their special physical properties [15–20]. The 2D materials are also used as coatings to form protective layers, which can act as physical barriers to prevent contact between the corrosive medium and the substrate and avoid corrosion [21,22]. Since graphene was discovered, it has been widely used in the field of corrosion resistance due to its good chemical stability, high strength, and low friction coefficient [23–25]. Single-layer defect-free graphene can prevent the permeation of molecules. However, defective graphene will accelerate galvanic corrosion [26,27]. In recent years, an increasing number of graphene-like 2D materials have been discovered, such as MoS_2, TiO_2, boron nitride, C_3N_4, and MXene [28–32]. TiO_2 and MoS_2 are two promising two-dimensional materials that have great development potential in the field of anticorrosion. TiO_2 has good chemical and thermal stability. At the same time, due to the lack of interconnected pores in TiO_2, the corrosion resistance of TiO_2 is higher than that of other metal oxides [33–37]. For instance, Devikala et al. [38] showed that with the increase in TiO_2 concentration the corrosion resistance efficiency of composites increased. Rostami et al. [39] demonstrated that the addition of TiO_2 improved the corrosion resistance of the pure cobalt film. Shams Anwar et al. [40] indicated that the addition of TiO_2 nanoparticles improved the corrosion resistance of Zn-Ni alloys. MoS_2 has attracted extensive attention in the field of corrosion protection because of its special (S-Mo-S) three-atomic-layer structure and excellent chemical stability [41,42]. For example, Xia et al. [43] prepared SiO_2 nanoparticles modified by MoS_2 nanosheets, which showed robust corrosion resistance. Hu et al. [44] demonstrated that nano-MoS_2 on a zinc phosphate coating effectively promoted the phosphating process, which improved the corrosion resistance of Q235 low-carbon steel. Chen et al. [45] loaded nano-MoS_2 on the surface of graphene oxide flakes, which endowed the composite coating with excellent barrier properties and significantly improved the corrosion resistance of the coating.

In this work, we describe our recent progress in the synthesis of the hydrophobic nanocomposite coating, which is 2D TiO_2/MoS_2 modified by a silane coupling agent, KH570. This was synthesized to enhance the corrosion resistance of AZ31B magnesium alloy. Because 2D TiO_2 and MoS_2 have similar electronegativity, MoS_2 was added to TiO_2 dispersion to obtain an electrophoretic solution. Then, a TiO_2/MoS_2 thin film was prepared on the surface of the magnesium alloy by electrophoretic deposition. The prepared 2D TiO_2/MoS_2 thin film was modified by KH570. The as-prepared thin film has hydrophobic properties and excellent corrosion resistance, which satisfies the demands of society and has broad application prospects.

2. Materials and Methods

2.1. Materials

AZ31B magnesium alloy (thickness of 1 mm, composition: 2.5–3.5 wt % Al, 0.6–1.4 wt % Zn, 0.2–1.0 wt % Mn, 0.04 wt % Ca, 0.003 wt % Fe, 0.001 wt % Ni, 0.08 wt % Si, 0.01 wt % Cu) was purchased from Dongguan Hongdi Metal Materials Co., Ltd. (Dongguan, China). Silane coupling agent KH570 was purchased from Jinan Xingfeilong Chemical Co., Ltd. (Jinan, China).

2.2. Preparation of TiO_2/MoS_2 Films

The TiO_2 colloid and MoS_2 were synthesized according to previous studies [46,47]. We added 0.7 mL of 1.0 mg/mL MoS_2 aqueous solution to 100 mL of 70% ethanol solution containing 1 mL of TiO_2 colloid. The two magnesium alloy substrates were kept at a distance of 15–20 mm in 5 mL of electrophoresis solution and electrophoresed for 90 s at a constant voltage of 20 V. After electrophoresis, the excess precipitates on the surface were washed with deionized water, dried at room temperature for 48 h, and heated at 150 °C for 60 min to obtain TiO_2/MoS_2 coating.

2.3. Preparation of TiO_2/MoS_2 Films Modified by KH570

The prepared samples were immersed in a beaker containing silane coupling agent, KH570 solution, for 30 min. After repeated rinsing with anhydrous ethanol, the silane-modified TiO_2/MoS_2 surface layer was obtained. The resultant product was denoted as TiO_2/MoS_2-K.

2.4. Analysis of Zeta Potential

The zeta potential of TiO_2 and MoS_2 was determined by Nanometrics (Malvern Zetasizer Nano ZS90, malvernpanalytical). Figure 1 shows the zeta potential diagrams of TiO_2 and MoS_2 measured at 25 °C. The measured electronegativity results are shown in Table 1. On the basis of Figure 1 and Table 1, it is shown that TiO_2 and MoS_2 have similar electronegativity. Therefore, MoS_2 was added to the TiO_2 dispersion, and both were deposited on the anode during electrophoresis deposition.

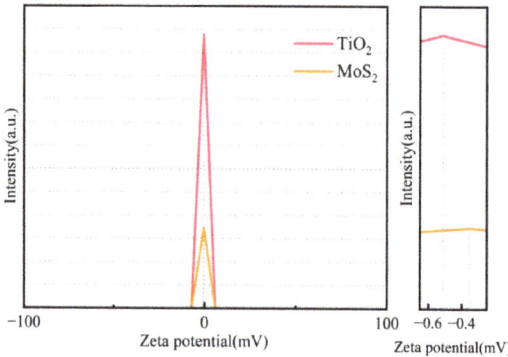

Figure 1. Zeta potential diagram of TiO_2 and MoS_2.

Table 1. Electronegativity of TiO_2 and MoS_2.

Sample	Zeta Potential/mV
TiO_2	−0.511
MoS_2	−0.356

2.5. Characteristics of Morphology, Chemical Composition, and Hydrophobic and Anticorrosion Properties

The surface morphology of the samples was observed using an S-4800 scanning electron microscope (SEM, S-4800, Japan Hitachi Corporation, Tokyo, Japan). The phase structure of the samples was characterized by a D-8 X-ray diffractometer (XRD, D8, Beijing Brook Technology Co., Ltd., Beijing, China). The X-ray photoelectron spectroscopy (XPS) analysis was performed using a Thermo ESCALAB 250XI. Fourier transform infrared spectroscopy (FTIR, IRPrestige-21, Shanghai Yixiang Instrument Co., Ltd., Shanghai, China) was carried out for TiO_2/MoS_2 and TiO_2/MoS_2-K. Measurement of the static water contact angle on the sample surface was performed using a contact angle measuring instrument (SDC-100S, Dongguan Shengding Precision Instrument Co., Ltd., Dongguan, China); 6 μL was set as the drip volume in the test, and multiple measurements were performed at different positions on the surface of each sample, with the average value of multiple measurements taken for analysis. Metal corrosion in the atmosphere was simulated by a salt spray corrosion test box (DF-YWX/Q-150, Nanjing Defu Test Equipment Co., Ltd., Nanjing, China); 5 wt % NaCl solution was prepared as the test solution. The experimental temperature was controlled at 35 ± 2 °C, and the pressure of the atomized salt solution was maintained in the range of 69~172 kPa. The sample size was 10 × 10 × 1 mm. An electrochemical workstation (CHI760E, Shanghai Chenhua, Shanghai, China) was used to test the corrosion polarization curve of the magnesium alloy samples. A 3.5% NaCl solution was used as the electrolyte, and a three-electrode system was adopted, in which AgCl was used as the reference electrode and Pt was used as the counter electrode, and the scanning speed was 5 mV/s. The determination of E_{corr} is the abscissa corresponding to the intersection of the cathode zone and the anode zone, and I_{coor} is the intersection of the tangent of the cathode zone and the vertical line corresponding to E_{corr}.

3. Results and Discussion

3.1. Morphology Analysis of TiO$_2$/MoS$_2$ Films

The surface morphology of the films deposited on the magnesium alloy substrate at different magnifications is presented in Figure 2. According to Figure 2a,b, a layer of a fog-like film composed of nanosheets is evenly spread on the surface of AZ31B. The bright white point is due to the tiny residual particles of titanate when it is exfoliated into TiO$_2$. In Figure 2c,d, the cracks are clearly displayed, which are caused by the accumulation of large amounts of MoS$_2$ in the substrate surface solution during electrophoretic deposition. Figure 2d shows the microscopic magnification of the relatively flat area. The nanosheets are stacked together to form large particles. The surface roughness was increased, and obvious cracks could be observed. As shown in Figure 2e, the surface layer completely and uniformly covers the surface of AZ31B, and there is no honeycomb connection between the nanosheets, which becomes very dense. The contact between the surface of AZ31B and air or corrosive substances is completely isolated, and a good protective barrier is formed [48,49]. Figure 2f shows that the coating becomes denser after KH570 modification.

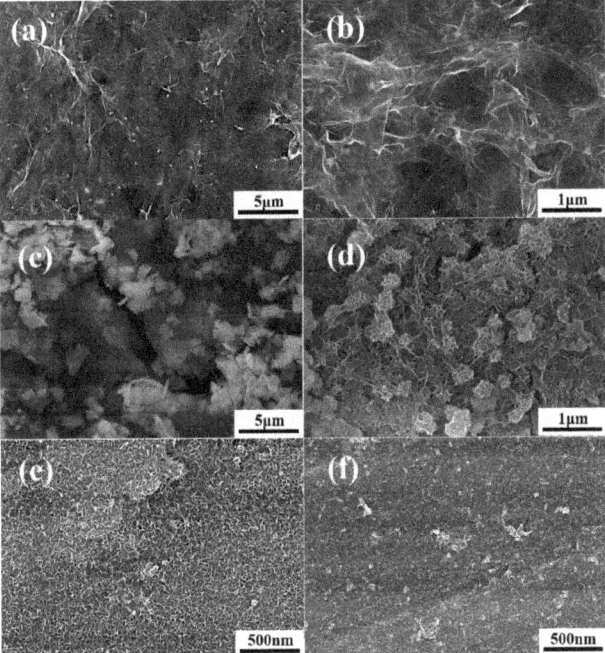

Figure 2. SEM images of films deposited on AZ31B: (**a**,**b**) TiO$_2$, (**c**,**d**) MoS$_2$, (**e**) TiO$_2$/MoS$_2$, (**f**) TiO$_2$/MoS$_2$-K.

3.2. Compositional Analysis of TiO$_2$/MoS$_2$ Films

Figure 3 shows the XRD patterns of the AZ31B substrate, TiO$_2$ and MoS$_2$ powder and TiO$_2$/MoS$_2$ film. The difference in peak position represents the difference in TiO$_2$ layer spacing, which proves that TiO$_2$ with random layer spacing is prepared [50,51]. The peaks of MoS$_2$ powder correspond to the hexagonal MoS$_2$ standard card (JCPDS No. 37-1492). Notably, after preparing TiO$_2$/MoS$_2$ film on the surface of the magnesium alloy, the peaks of TiO$_2$ and MoS$_2$ are not obvious due to the strong peak of the AZ31B substrate. Combined with the results of the energy spectrum analysis of the corrosion products of the samples, TiO_2/MoS_2 films were successfully prepared.

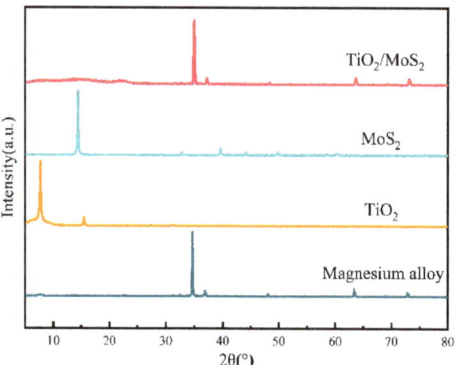

Figure 3. XRD patterns of magnesium alloy, TiO_2, MoS_2, and TiO_2/MoS_2 films.

3.3. XPS of TiO_2/MoS_2 Coating

XPS measurements were conducted to characterize the surface compositions. As shown in Figure 4, TiO_2 features two characteristic peaks at around 463.9 and 458.2 eV, corresponding to the Ti $2p^{1/2}$ and Ti $2p^{3/2}$ components, respectively. In Figure 4c, two peaks at 233.1 and 229.9 eV are assigned to Mo $3d^{3/2}$ and Mo $d^{5/2}$, which suggests that the majority of Mo at the surface is Mo^{4+}. Simultaneously, the peak at 227.1 eV belongs to the S 2s orbital of MoS_2. The S 2p XPS spectrum shows that two peaks located at 163.9 and 162.9 eV correspond to S $2p^{1/2}$ and S $2p^{3/2}$, which is consistent with the −2 oxidation state of sulfur. This confirms the successful synthesis of TiO_2/MoS_2.

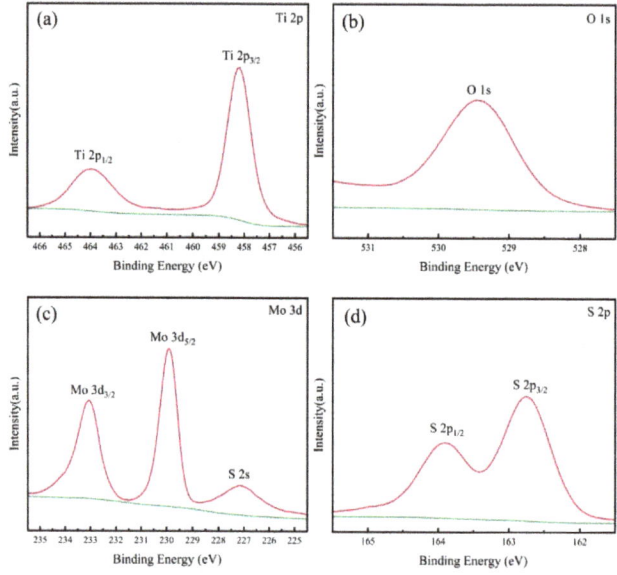

Figure 4. XPS spectra of TiO_2/MoS_2: (**a**) Ti 2p, (**b**) O 1s, (**c**) Mo 3d, (**d**) S 2p.

3.4. Hydrophilicity Test of TiO_2/MoS_2 Films

The wettability of the material surface also has an important influence on the corrosion resistance of the material. It is generally believed that the hydrophobic surface makes it more difficult for water molecules or chloride ions to penetrate the metal surface due to the isolation effect, thereby inhibiting the corrosion of the metal [52–54]. The contact angles (CA) of the magnesium alloy and the TiO_2/MoS_2 and TiO_2/MoS_2-K films are presented in Figure 5a. As shown in Figure 5a, the CA of the magnesium alloy is 36.08°, indicating that the surface of the AZ31B substrate without coatings is hydrophilic. When the TiO_2/MoS_2 surface layer was electrophoretically deposited on the surface of the AZ31B substrate, the CA increased to 65.43°. When the TiO_2/MoS_2 film was modified by KH570, the CA value was 131.53°. Another important observation is that the surface of the alloy changes from hydrophilic to hydrophobic. This is because the silane coupling agent KH570 is hydrolyzed into silanol in the solution, and the silanol is condensed with -OH on the TiO_2/MoS_2 film [55]. The hydrophilic functional group (-OH) of the TiO_2/MoS_2 coating was replaced by the organic functional group in silanol, and the reaction process is shown in Formulas (1)–(3). Figure 5b shows the FTIR spectra of TiO_2/MoS_2 and TiO_2/MoS_2-K. Two new peaks of TiO_2/MoS_2-K at about 1240 and 1280 cm^{-1} are attributed to stretching vibrations of C-O-C. Meanwhile, the absorption peak at 1710 cm^{-1} is assigned to the stretching vibration of C=O. The intensity of TiO_2/MoS_2-K is weaker than that of TiO_2/MoS_2 because the organic functional groups in silanol replace -OH on the surface of silane-modified TiO_2/MoS_2. These results indicate that the surface of the TiO_2/MoS_2 surface layer is modified by KH570, thereby enhancing the surface hydrophobicity of TiO_2/MoS_2 [56]. This effectively isolates the magnesium alloy from making contact with liquid, making it more difficult for water molecules or chloride ions to penetrate the metal surface and effectively enhancing the corrosion resistance of the magnesium alloy [57].

$$CH_2=C(CH_3)COO(CH_2)_3Si(OCH_3)_3 + 3H_2O \rightarrow CH_2=C(CH_3)COO(CH_2)_3Si(OH)_3 + 3CH_3OH \qquad (1)$$

$$CH_2=C(CH_3)COO(CH_2)_3Si(OH)_3 + HO\text{-}(TiO_2)_n \rightarrow CH_2=C(CH_3)COO(CH_2)_3Si\text{-}O\text{-}(TiO_2)_n + H_2O + 2OH^- \qquad (2)$$

$$CH_2=C(CH_3)COO(CH_2)_3Si(OH)_3 + HO\text{-}(MoS_2)_n \rightarrow CH_2=C(CH_3)COO(CH_2)_3Si\text{-}O\text{-}(MoS_2)_n + H_2O + 2OH^- \qquad (3)$$

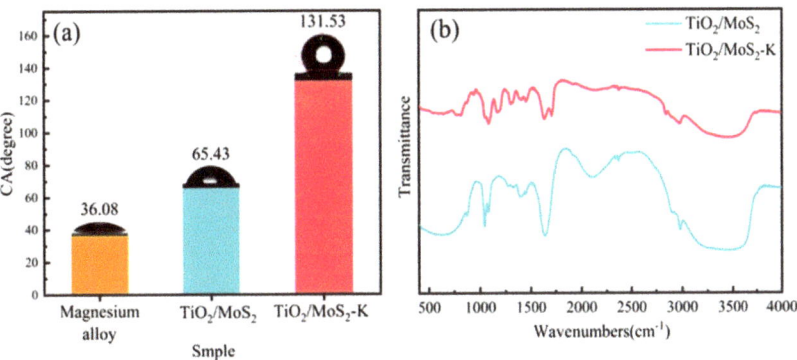

Figure 5. (**a**) Contact angles of magnesium alloy, TiO_2/MoS_2, and TiO_2/MoS_2-K; (**b**) FT-IR spectra of TiO_2/MoS_2 and TiO_2/MoS_2-K.

3.5. Electrochemical Test of TiO_2/MoS_2 Films

Figure 6 shows the potentiodynamic polarization curves of all the samples in 3.5 wt % NaCl aqueous solution. The results of the test are summarized in Table 2. On the basis of Figure 6 and Table 2, it is shown that the corrosion current density (I_{corr}) of the magnesium alloy decreases after electrophoresis. The polarization curve shape of the sample after deposition is similar to that of the magnesium alloy substrate, indicating that the 2D films do not change the corrosion kinetics process of the magnesium alloy. The contact between the substrate and the corrosion medium is isolated by the surface thin film, which improves the corrosion resistance of the magnesium alloy to a certain extent. The corrosion potential (E_{corr}) of the AZ31B magnesium alloy is -1.47 V and the I_{corr} is 6.81×10^{-4} A·cm^{-2}, indicating that the corrosion resistance of the magnesium alloy matrix is poor [58]. The results show that the E_{corr} of TiO_2/MoS_2-K has a higher potential compared with TiO_2, MoS_2, and TiO_2/MoS_2 composite coatings. Among them, TiO_2/MoS_2-K has the lowest I_{corr}. The I_{corr} values for the TiO_2 and MoS_2 surface layer are 5.13×10^{-6} and 2.19×10^{-5} A·cm^{-2}, respectively, whereas that of the TiO_2/MoS_2 film is 3.69×10^{-7} A·cm^{-2}. The I_{corr} of the TiO_2/MoS_2 film decreases by two orders of magnitude compared to the MoS_2 film, which is due to the dense state of the TiO_2/MoS_2 coating. The E_{corr} of the TiO_2/MoS_2-K film is -0.85 V, and the I_{corr} is 6.73×10^{-8} A·cm^{-2}. Compared with the TiO_2/MoS_2 film, the E_{corr} of the TiO_2/MoS_2-K film shifted positively by 0.15 V, and the I_{corr} decreased by six times. This indicates that the KH570 modification improved the corrosion resistance of the TiO_2/MoS_2 protective layer, which is ascribed to its effectively isolating the magnesium alloy from contacting the corrosive liquid.

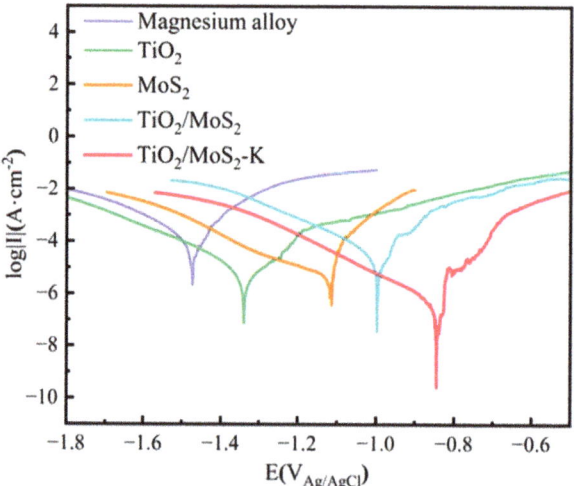

Figure 6. Polarization curves of magnesium alloy, TiO_2, MoS_2, TiO_2/MoS_2, TiO_2/MoS_2-K.

Table 2. Potentiodynamic polarization curves of samples in 3.5 wt % NaCl aqueous solution.

Sample	$E_{corr}/(V_{Ag/AgCl})$	$I_{corr}/(A·cm^{-2})$
Magnesium alloy	-1.47	6.81×10^{-4}
TiO_2	-1.34	5.31×10^{-6}
MoS_2	-1.11	2.19×10^{-5}
TiO_2/MoS_2	-1.00	3.69×10^{-7}
TiO_2/MoS_2-K	-0.85	6.73×10^{-8}

To further characterize corrosion resistance, we used EIS to analyze the anticorrosion of the sample. Figure 7 shows the electrochemical impedance spectroscopy of magnesium alloy, TiO_2, TiO_2/MoS_2-K. The electrochemical impedance spectroscopy data are described in detail by the numerical fitting of experimental data. The corresponding equivalent circuit was selected to fit the impedance data, as shown in Figure 8. The fitting circuit of the magnesium alloy substrate is shown in Figure 8a, and the fitting circuit of the samples with TiO_2 film and those with TiO_2/MoS_2-K films is shown in Figure 8b [59]. The results are shown in Table 3. The circuit involves the solution resistance (R_s), the charge transfer resistance (R_{ct}) of the Mg particles, and the electric double-layer capacitance (CPE_{dl}) at the interface between the electrolyte and magnesium particles. R_p is the inductance resistance, and CPL_{film} is the inductance, corresponding to the electrode reaction between the film layer and the electrolyte interface. It is reported that R_{ct} is closely related to the corrosion process, that is, the higher the R_{ct}, the better the corrosion resistance. It can be seen from the results that the impedance radius of the TiO_2/MoS_2-K sample is the largest, and the R_{ct} value is the largest at 871.9 $\Omega \cdot cm^2$, which is significantly higher than that of the magnesium alloy sample and the sample with the TiO_2 film. It shows that the corrosion resistance of TiO_2/MoS_2-K is the highest, which is consistent with the results of the polarization curve.

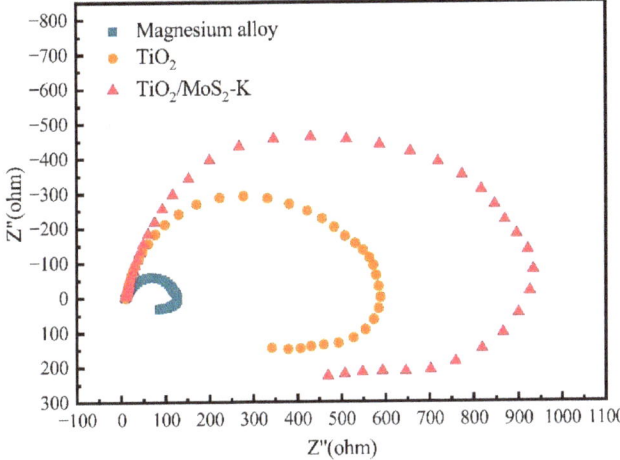

Figure 7. Electrochemical impedance spectroscopy of magnesium alloy, TiO_2, TiO_2/MoS_2-K.

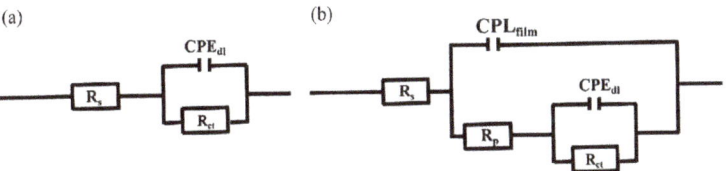

Figure 8. Equivalent circuits for EIS spectra: (**a**) magnesium alloy, (**b**) TiO_2 thin film and TiO_2/MoS_2-K thin film.

Table 3. Electrochemical data extracted from ECs fitting of the EIS curves.

Sample	R_s ($\Omega \cdot cm^2$)	CPE_{film} (F/cm^2)	R_p ($\Omega \cdot cm^2$)	CPE_{dl}-T (F/cm^2)	CPE_{dl}-P (F/cm^2)	R_{ct} ($\Omega \cdot cm^2$)
Magnesium alloy	10.2	-	-	7.11×10^{-6}	-	111.9
TiO_2	8.97	1.27×10^{-6}	16.96	1.66×10^{-6}	0.98	659.6
TiO_2/MoS_2-K	9.25	8.92×10^{-6}	19.29	1.87×10^{-6}	0.95	871.9

3.6. Salt Spray Corrosion Experiment of TiO_2/MoS_2 Films

Figure 9 shows the comparison of the surface morphology of the magnesium alloy samples, electrophoretically deposited TiO_2, and TiO_2/MoS_2-K after the neutral salt spray test for 24, 48, and 72 h. Table 4 shows the results of the energy spectrum analysis of the corrosion products of the samples. It can be seen from the data in Table 4 that, in addition to the original elements, Na and Cl elements were also added to all the corrosion products of the samples, indicating that the chemical corrosion process occurred in the samples and salt spray. Combined with the SEM images, it can be seen that pitting corrosion occurred on the samples with TiO_2/MoS_2-K film, which is similar to that on the samples with TiO_2 coating. With increasing time, the corrosion area increased gradually, but it was smaller than that on the surface of the magnesium alloy and the samples with the TiO_2 layer. This was mainly due to the good chemical stability of TiO_2 and MoS_2 and the formation of the dense protective layer after electrophoresis. The hydrophobic thin film modified by KH570 further delayed the corrosion of the magnesium alloy.

Figure 9. The microstructure of the sample surface after salt spray test at different times. Magnesium alloy: (**a**) 24 h, (**b**) 48 h, (**c**) 72 h; TiO_2 film: (**e**) 24 h, (**f**) 48 h, (**g**) 72 h; TiO_2/MoS_2-K film: (**i**) 24 h, (**j**) 48 h, (**k**) 72 h. The (**d**,**h**,**l**) insets are partial enlargements of (**c**,**g**,**k**), respectively.

Table 4. Energy spectrum analysis of neutral salt spray test products (at. %).

Point	O	Mg	Mn	Ti	Na	Cl	Mo	S
A	57.64	26.18	12.02	0	2.79	1.37	0	0
B	59.23	25.43	11.08	0	2.51	1.75	0	0
C	55.38	25.28	12.60	2.40	3.15	1.19	0	0
D	53.85	25.32	11.06	3.99	3.98	1.80	0	0
E	55.62	25.16	11.10	3.69	2.17	1.12	0.22	0.92
F	56.38	25.09	11.20	2.57	2.23	1.33	0.19	1.01

Figure 10 shows the salt spray corrosion weight gain of the magnesium alloy samples after electrophoretic deposition of TiO_2 and TiO_2/MoS_2-K films. The weight gain rate of the sample with TiO_2/MoS_2-K is higher at the beginning of the test than that of the magnesium alloy sample and the sample with TiO_2 coating. With increasing salt spray corrosion time, the weight gain rate of the sample gradually slows down because the corrosion products on the surface of the sample increase, which hinders the continuous reaction of chloride ion contact with the magnesium alloy surface. In addition, the weight gain of the sample with the TiO_2/MoS_2-K film is significantly lower than that of the former two samples, and the corrosion rate is the lowest. This result indicates that the TiO_2/MoS_2-K protective layer delayed the corrosion of the magnesium alloy to a greater extent and had the best corrosion resistance.

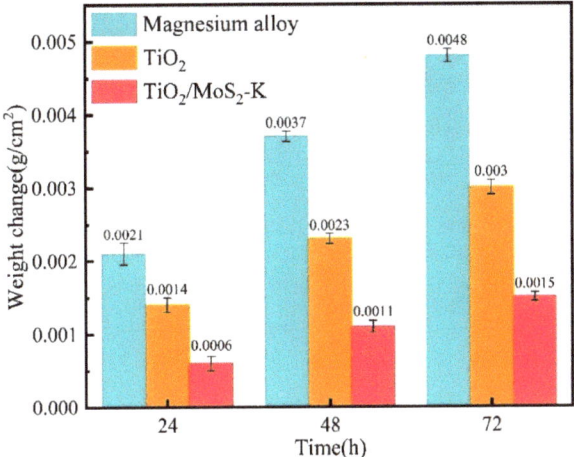

Figure 10. Salt spray corrosion weight gain diagram.

4. Conclusions

The TiO$_2$/MoS$_2$ coating was successfully prepared on the surface of the magnesium alloy by adding MoS$_2$ in the electrophoretic deposition of TiO$_2$. After the TiO$_2$/MoS$_2$ layer was modified by the silane coupling agent KH570, the wettability of the magnesium alloy surface was changed from hydrophilic to hydrophobic. The as-prepared TiO$_2$/MoS$_2$-K film had a contact angle of 131.53°. Combined with the electrochemical test and salt spray corrosion analysis, it can be seen that the TiO$_2$/MoS$_2$-K film developed a significantly improved anticorrosion property compared to the magnesium alloy. In addition, we anticipate a promising potential to transfer this technology to other metal substrates for important applications.

Author Contributions: Conceptualization, Q.L. and Z.L.; methodology, H.W.; investigation, L.L.; writing-original draft preparation, L.L.; writing-review and editing, G.M.; supervision, Q.L.; project administration, L.Z.; funding acquisition, Q.L. All authors have read and agreed to the published version of the manuscript.

Funding: This work was supported by the Natural Science Foundation of Anhui Province (No. 2008085ME132) and the Key Project of Anhui Provincial Department of Education (No. KJ2019A0157).

Conflicts of Interest: The authors declare no conflict of interest.

References

1. Luo, K.; Zhang, L.; Wu, G.; Liu, W.; Ding, W. Effect of Y and Gd content on the microstructure and mechanical properties of Mg–Y–RE alloys. *J. Magnes. Alloy.* **2019**, *7*, 345–354. [CrossRef]
2. Yang, Y.; Xiong, X.; Chen, J.; Peng, X.; Chen, D.; Pan, F. Research advances in magnesium and magnesium alloys worldwide in 2020. *J. Magnes. Alloy.* **2021**, *9*, 705–747. [CrossRef]
3. Liu, H.; Tong, Z.; Yang, Y.; Zhou, W.; Chen, J.; Pan, X.; Ren, X. Preparation of phosphate conversion coating on laser surface textured surface to improve corrosion performance of magnesium alloy. *J. Alloys Compd.* **2021**, *865*, 158701. [CrossRef]
4. Xianhua, C.; Yuxiao, G.; Fusheng, P. Research Progress in Magnesium Alloys as Functional Materials. *Rare Met. Mater. Eng.* **2016**, *45*, 2269–2274. [CrossRef]
5. Song, J.; She, J.; Chen, D.; Pan, F. Latest research advances on magnesium and magnesium alloys worldwide. *J. Magnes. Alloy.* **2020**, *8*, 1–41. [CrossRef]
6. Siddique, S.; Bernussi, A.A.; Husain, S.W.; Yasir, M. Enhancing structural integrity, corrosion resistance and wear properties of Mg alloy by heat treated cold sprayed Al coating. *Surf. Coat. Technol.* **2020**, *394*, 125882. [CrossRef]
7. Yeganeh, M.; Mohammadi, N. Superhydrophobic surface of Mg alloys: A review. *J. Magnes. Alloy.* **2018**, *6*, 59–70. [CrossRef]
8. Chang, S.-H.; Niu, L.; Su, Y.; Wang, X.; Tong, X.; Li, G. Effect of the pretreatment of silicone penetrant on the performance of the chromium-free chemfilm coated on AZ91D magnesium alloys. *Mater. Chem. Phys.* **2016**, *171*, 312–317. [CrossRef]

9. Ballam, L.R.; Arab, H.; Bestetti, M.; Franz, S.; Masi, G.; Sola, R.; Donati, L.; Martini, C. Improving the Corrosion Resistance of Wrought ZM21 Magnesium Alloys by Plasma Electrolytic Oxidation and Powder Coating. *Materials* **2021**, *14*, 2268. [CrossRef]
10. Liu, C.; Liang, J.; Zhou, J.; Wang, L.; Li, Q. Effect of laser surface melting on microstructure and corrosion characteristics of AM60B magnesium alloy. *Appl. Surf. Sci.* **2015**, *343*, 133–140. [CrossRef]
11. Lu, F.-F.; Ma, K.; Li, C.-X.; Yasir, M.; Luo, X.-T.; Li, C.-J. Enhanced corrosion resistance of cold-sprayed and shot-peened aluminum coatings on LA43M magnesium alloy. *Surf. Coat. Technol.* **2020**, *394*, 125865. [CrossRef]
12. Yao, H.-L.; Yi, Z.-H.; Yao, C.; Zhang, M.-X.; Wang, H.-T.; Li, S.-B.; Bai, X.-B.; Chen, Q.-Y.; Ji, G.-C. Improved corrosion resistance of AZ91D magnesium alloy coated by novel cold-sprayed Zn-HA/Zn double-layer coatings. *Ceram. Int.* **2020**, *46*, 7687–7693. [CrossRef]
13. Hosseini, M.R.; Ahangari, M.; Johar, M.H.; Allahkaram, S.R. Optimization of nano HA-SiC coating on AISI 316L medical grade stainless steel via electrophoretic deposition. *Mater. Lett.* **2021**, *285*, 129097. [CrossRef]
14. Guan, S.; Hao, L.; Yoshida, H.; Itoi, T.; Cheng, Y.; Seki, S.; Nishina, Y.; Lu, Y. Enhanced photocatalytic activity and stability of TiO2/graphene oxide composites coatings by electrophoresis deposition. *Mater. Lett.* **2021**, *286*, 129258. [CrossRef]
15. Sun, T.-Y.; Hao, Y.; Wu, Y.-H.; Zhao, W.-J.; Huang, L.-F. Corrosion Resistance of Ultrathin Two-Dimensional Coatings: First-Principles Calculations towards In-Depth Mechanism Understanding and Precise Material Design. *Metals* **2021**, *11*, 2011. [CrossRef]
16. Mak, K.F.; Shan, J. Photonics and optoelectronics of 2D semiconductor transition metal dichalcogenides. *Nat. Photonics* **2016**, *10*, 216–226. [CrossRef]
17. Lu, Q.; Yu, Y.; Ma, Q.; Chen, B.; Zhang, H. 2D Transition-Metal-Dichalcogenide-Nanosheet-Based Composites for Photocatalytic and Electrocatalytic Hydrogen Evolution Reactions. *Adv. Mater.* **2016**, *28*, 1917–1933. [CrossRef] [PubMed]
18. Kumar, K.S.; Choudhary, N.; Jung, Y.; Thomas, J. Recent Advances in Two-Dimensional Nanomaterials for Supercapacitor Electrode Applications. *ACS Energy Lett.* **2018**, *3*, 482–495. [CrossRef]
19. Choi, W.; Choudhary, N.; Han, G.H.; Park, J.; Akinwande, D.; Lee, Y.H. Recent development of two-dimensional transition metal dichalcogenides and their applications. *Mater. Today* **2017**, *20*, 116–130. [CrossRef]
20. Bhandavat, R.; David, L.; Singh, G. Synthesis of Surface-Functionalized WS2 Nanosheets and Performance as Li-Ion Battery Anodes. *J. Phys. Chem. Lett.* **2012**, *3*, 1523–1530. [CrossRef]
21. Ding, R.; Chen, S.; Lv, J.; Zhang, W.; Zhao, X.-d.; Liu, J.; Wang, X.; Gui, T.-J.; Li, B.-J.; Tang, Y.-Z.; et al. Study on graphene modified organic anti-corrosion coatings: A comprehensive review. *J. Alloys Compd.* **2019**, *806*, 611–635. [CrossRef]
22. Seel, M.; Pandey, R. Proton and hydrogen transport through two-dimensional monolayers. *2D Mater.* **2016**, *3*, 025004. [CrossRef]
23. Prasai, D.; Tuberquia, J.C.; Harl, R.R.; Jennings, G.K.; Bolotin, K.I. Graphene: Corrosion-inhibiting coating. *ACS Nano* **2012**, *6*, 1102–1108. [CrossRef] [PubMed]
24. Zhao, Z.; Hou, T.; Wu, N.; Jiao, S.; Zhou, K.; Yin, J.; Suk, J.W.; Cui, X.; Zhang, M.; Li, S.; et al. Polycrystalline Few-Layer Graphene as a Durable Anticorrosion Film for Copper. *Nano Lett.* **2021**, *21*, 1161–1168. [CrossRef] [PubMed]
25. Bohm, S. Graphene against corrosion. *Nat. Nanotechnol.* **2014**, *9*, 741–742. [CrossRef] [PubMed]
26. Cui, C.; Lim, A.T.O.; Huang, J. A cautionary note on graphene anti-corrosion coatings. *Nat. Nanotechnol.* **2017**, *12*, 834–835. [CrossRef] [PubMed]
27. Zhang, Y.; Sun, J.; Xiao, X.; Wang, N.; Meng, G.; Gu, L. Graphene-like two-dimensional nanosheets-based anticorrosive coatings: A review. *J. Mater. Sci. Technol.* **2022**, *129*, 139–162. [CrossRef]
28. Nurdiwijayanto, L.; Nishijima, H.; Miyake, Y.; Sakai, N.; Osada, M.; Sasaki, T.; Taniguchi, T. Solution-Processed Two-Dimensional Metal Oxide Anticorrosion Nanocoating. *Nano Lett.* **2021**, *21*, 7044–7049. [CrossRef]
29. Shi, K.; Meng, X.; Xiao, S.; Chen, G.; Wu, H.; Zhou, C.; Jiang, S.; Chu, P.K. MXene Coatings: Novel Hydrogen Permeation Barriers for Pipe Steels. *Nanomaterials* **2021**, *11*, 2737. [CrossRef]
30. Mujib, S.B.; Mukherjee, S.; Ren, Z.; Singh, G. Assessing corrosion resistance of two-dimensional nanomaterial-based coatings on stainless steel substrates. *R. Soc. Open Sci.* **2020**, *7*, 200214. [CrossRef]
31. Shen, L.; Zhao, W.; Wang, K.; Xu, J. GO-Ti$_3$C$_2$ two-dimensional heterojunction nanomaterial for anticorrosion enhancement of epoxy zinc-rich coatings. *J. Hazard. Mater.* **2021**, *417*, 126048. [CrossRef] [PubMed]
32. Xi, K.; Wu, H.; Zhou, C.; Qi, Z.; Yang, K.; Fu, R.K.Y.; Xiao, S.; Wu, G.; Ding, K.; Chen, G.; et al. Improved corrosion and wear resistance of micro-arc oxidation coatings on the 2024 aluminum alloy by incorporation of quasi-two-dimensional sericite microplates. *Appl. Surf. Sci.* **2022**, *585*, 152693. [CrossRef]
33. Wang, X. Preparation and Corrosion Resistance of AKT-Waterborne Polyurethane Coating. *Int. J. Electrochem. Sci.* **2020**, *15*, 1450–1464. [CrossRef]
34. Kavimani, V.; Prakash, K.S.; Gunashri, R.; Sathish, P. Corrosion protection behaviour of r-GO/TiO$_2$ hybrid composite coating on Magnesium substrate in 3.5 wt.% NaCl. *Prog. Org. Coat.* **2018**, *125*, 358–364. [CrossRef]
35. Li, Z.; Ding, S.; Kong, L.; Wang, X.; Ashour, A.; Han, B.; Ou, J. Nano TiO$_2$-engineered anti-corrosion concrete for sewage system. *J. Clean. Prod.* **2022**, *337*, 130508. [CrossRef]
36. Kumar, A.M.; Khan, A.; Hussein, M.A.; Khan, M.Y.; Dafalla, H.; Suresh, B.; Ramakrishna, S. Hybrid nanocomposite coatings from PEDOT and BN-TiO$_2$ nanosheets: Enhanced invitro corrosion resistance, wettability and biocompatibility for biomedical applications. *Prog. Org. Coat.* **2022**, *170*, 106946. [CrossRef]

37. Zhang, Y.; Zhang, K.; Lei, S.; Su, Y.; Yang, W.; Wang, J.; Qin, G.; Li, W. Formation and oxidation behavior of TiO_2 modified Al_2O_3-Nb_2O_5/$NbAl_3$ composite coating prepared by two-step methods. *Surf. Coat. Technol.* **2022**, *433*, 128081. [CrossRef]
38. Devikala, S.; Kamaraj, P.; Arthanareeswari, M. Corrosion resistance behavior of PVA/TiO_2 composite in 3.5% NaCl. *Mater. Today Proc.* **2018**, *5*, 8672–8677. [CrossRef]
39. Rostami, S.; Mahdavi, S.; Alinezhadfar, M.; Mohseni, A. Tribological and corrosion behavior of electrochemically deposited Co/TiO_2 micro/nano-composite coatings. *Surf. Coat. Technol.* **2021**, *423*, 127591. [CrossRef]
40. Anwar, S.; Khan, F.; Zhang, Y. Corrosion behaviour of Zn-Ni alloy and Zn-Ni-nano-TiO_2 composite coatings electrodeposited from ammonium citrate baths. *Process. Saf. Environ. Prot.* **2020**, *141*, 366–379. [CrossRef]
41. Wang, K.; Wang, J.; Fan, J.; Lotya, M.; O'Neill, A.; Fox, D.; Feng, Y.; Zhang, X.; Jiang, B.; Zhao, Q.; et al. Ultrafast saturable absorption of two-dimensional MoS_2 nanosheets. *ACS Nano* **2013**, *7*, 9260–9267. [CrossRef]
42. Asan, G.; Asan, A.; Çelikkan, H. The effect of 2D-MoS_2 doped polypyrrole coatings on brass corrosion. *J. Mol. Struct.* **2020**, *1203*, 127318. [CrossRef]
43. Xia, Y.; He, Y.; Chen, C.; Wu, Y.; Chen, J. MoS_2 nanosheets modified SiO_2 to enhance the anticorrosive and mechanical performance of epoxy coating. *Prog. Org. Coat.* **2019**, *132*, 316–327. [CrossRef]
44. Hu, S.; Muhammad, M.; Wang, M.; Ma, R.; Du, A.; Fan, Y.; Cao, X.; Zhao, X. Corrosion resistance performance of nano-MoS_2-containing zinc phosphate coating on Q235 steel. *Mater. Lett.* **2020**, *265*, 127256. [CrossRef]
45. Chen, C.; He, Y.; Xiao, G.; Xia, Y.; Li, H.; He, Z. Two-dimensional hybrid materials: MoS_2-RGO nanocomposites enhanced the barrier properties of epoxy coating. *Appl. Surf. Sci.* **2018**, *444*, 511–521. [CrossRef]
46. Sasaki, T.; Kooli, F.; Iida, M.; Michiue, Y.; Takenouchi, S.; Yajima, Y.; Izumi, F.; Chakoumakos, B.C.; Watanabe, M. A Mixed Alkali Metal Titanate with the Lepidocrocite-like Layered Structure. Preparation, Crystal Structure, Protonic Form, and Acid−Base Intercalation Properties. *Chem. Mater.* **1998**, *10*, 4123–4128. [CrossRef]
47. Zhang, Z.; Li, W.; Yuen, M.F.; Ng, T.-W.; Tang, Y.; Lee, C.-S.; Chen, X.; Zhang, W. Hierarchical composite structure of few-layers MoS_2 nanosheets supported by vertical graphene on carbon cloth for high-performance hydrogen evolution reaction. *Nano Energy* **2015**, *18*, 196–204. [CrossRef]
48. Singh, S.; Singh, G.; Bala, N. Corrosion behavior and characterization of HA/Fe_3O_4/CS composite coatings on AZ91 Mg alloy by electrophoretic deposition. *Mater. Chem. Phys.* **2019**, *237*, 121884. [CrossRef]
49. Hu, S.; Li, W.; Finklea, H.; Liu, X. A review of electrophoretic deposition of metal oxides and its application in solid oxide fuel cells. *Adv. Colloid Interface Sci.* **2020**, *276*, 102102. [CrossRef] [PubMed]
50. Hoshide, T.; Zheng, Y.; Hou, J.; Wang, Z.; Li, Q.; Zhao, Z.; Ma, R.; Sasaki, T.; Geng, F. Flexible Lithium-Ion Fiber Battery by the Regular Stacking of Two-Dimensional Titanium Oxide Nanosheets Hybridized with Reduced Graphene Oxide. *Nano Lett.* **2017**, *17*, 3543–3549. [CrossRef]
51. Maluangnont, T.; Matsuba, K.; Geng, F.; Ma, R.; Yamauchi, Y.; Sasaki, T. Osmotic Swelling of Layered Compounds as a Route to Producing High-Quality Two-Dimensional Materials. A Comparative Study of Tetramethylammonium versus Tetrabutylammonium Cation in a Lepidocrocite-type Titanate. *Chem. Mater.* **2013**, *25*, 3137–3146. [CrossRef]
52. Peng, F.; Zhang, D.; Liu, X.; Zhang, Y. Recent progress in superhydrophobic coating on Mg alloys: A general review. *J. Magnes. Alloy.* **2021**, *9*, 1471–1486. [CrossRef]
53. Zhang, M.; Xu, H.; Zeze, A.L.P.; Liu, X.; Tao, M. Coating performance, durability and anti-corrosion mechanism of organic modified geopolymer composite for marine concrete protection. *Cem. Concr. Compos.* **2022**, *129*, 104495. [CrossRef]
54. Gong, C.; Jianzhong, L.; Cuicui, C.; Changfeng, L.; Liang, S. Study on silane impregnation for protection of high performance concrete. *Procedia Eng.* **2012**, *27*, 301–307. [CrossRef]
55. Wang, S.; Wang, Y.; Zou, Y.; Wu, Y.; Chen, G.; Ouyang, J.; Jia, D.; Zhou, Y. A self-adjusting PTFE/TiO_2 hydrophobic double-layer coating for corrosion resistance and electrical insulation. *Chem. Eng. J.* **2020**, *402*, 126116. [CrossRef]
56. Parichehr, R.; Dehghanian, C.; Nikbakht, A. Preparation of PEO/silane composite coating on AZ31 magnesium alloy and investigation of its properties. *J. Alloys Compd.* **2021**, *876*, 159995. [CrossRef]
57. Wu, C.; Liu, Q.; Chen, R.; Liu, J.; Zhang, H.; Li, R.; Takahashi, K.; Liu, P.; Wang, J. Fabrication of ZIF-8@SiO_2 Micro/Nano Hierarchical Superhydrophobic Surface on AZ31 Magnesium Alloy with Impressive Corrosion Resistance and Abrasion Resistance. *ACS Appl. Mater. Interfaces* **2017**, *9*, 11106–11115. [CrossRef]
58. Zhang, G.; Qin, S.; Yan, L.; Zhang, X. Simultaneous improvement of electromagnetic shielding effectiveness and corrosion resistance in magnesium alloys by electropulsing. *Mater. Charact.* **2021**, *174*, 111042. [CrossRef]
59. Liu, H.; Tong, Z.; Zhou, W.; Yang, Y.; Jiao, J.; Ren, X. Improving electrochemical corrosion properties of AZ31 magnesium alloy via phosphate conversion with laser shock peening pretreatment. *J. Alloys Compd.* **2020**, *846*, 155837. [CrossRef]

Article

Effect of the Testing Temperature on the Impact Property of a Multilayered Soft–Hard Copper–Brass Block

Tong Liu [1,2], Jiansheng Li [1,2,*], Kezhang Liu [1], Mengmeng Wang [1,2], Yu Zhao [1,2,3,*], Zhongchen Zhou [4], Yong Wei [1], Qi Yang [1], Ming Chen [1], Qingzhong Mao [4,*], and Yufeng Sun [1,2]

[1] School of Materials Science and Engineering, Anhui Polytechnic University, Wuhu 241000, China
[2] Anhui Key Laboratory of High-Performance Non-Ferrous Metal Materials, Anhui Polytechnic University, Wuhu 241000, China
[3] NBTM New Materials Group Corporation Limited, Ningbo 315000, China
[4] School of Materials Science and Engineering, Nanjing University of Science and Technology, Nanjing 210094, China
* Correspondence: lijiansheng@ahpu.edu.cn or drlijiansheng@163.com (J.L.); zhaoyu@ahpu.edu.cn (Y.Z.); 216116000150@njust.edu.cn (Q.M.)

Citation: Liu, T.; Li, J.; Liu, K.; Wang, M.; Zhao, Y.; Zhou, Z.; Wei, Y.; Yang, Q.; Chen, M.; Mao, Q.; et al. Effect of the Testing Temperature on the Impact Property of a Multilayered Soft–Hard Copper–Brass Block. *Coatings* **2022**, *12*, 1236. https://doi.org/10.3390/coatings12091236

Academic Editor: Alexandru Enesca

Received: 29 July 2022
Accepted: 20 August 2022
Published: 24 August 2022

Publisher's Note: MDPI stays neutral with regard to jurisdictional claims in published maps and institutional affiliations.

Copyright: © 2022 by the authors. Licensee MDPI, Basel, Switzerland. This article is an open access article distributed under the terms and conditions of the Creative Commons Attribution (CC BY) license (https://creativecommons.org/licenses/by/4.0/).

Abstract: The impact property is one of the most significant mechanical properties for metallic materials. In the current work, a soft–hard copper–brass block with a high yield strength of ~320 MPa and good uniform elongation of ~20% was prepared, and the effect of the testing temperature on its impact property was explored. The results showed that the impact energy was decreased with the increase in testing temperature. The impact energies at liquid nitrogen temperature (LNT), room temperature (RT), and 200 °C were 8.15 J, 7.39 J, and 7.04 J, respectively. The highest impact energy at LNT was attributed to the coordinated plastic deformation effects, which was indicated by the tiny dimples during the process of the delamination of soft–hard copper–brass interfaces. The high temperature of 200 °C can weaken the copper–brass interface and reduce the absorption of deformation energy, result in low impact energy.

Keywords: copper–brass block; testing temperature; impact property; multilayered structure; delamination

1. Introduction

Pure copper and copper alloys are widely applied in thermal conductive devices and electronic equipment because of their superior thermal and electrical properties [1,2]. However, the low yield strength of pure copper and its alloys, especially for their coarse-grained states, may limit their application as structural parts in thermal and electrical applications [3–5]. Severe plastic deformation can turn the coarse-grained structures into ultrafine or nano-grained structures, which will greatly enhance the strength of pure copper and its alloys [6–8]. However, the toughness was dramatically reduced by the defects formed in the deformed structures, and thus increased the risk of catastrophic failure of mechanical parts [5,9]. Fabricating/processing copper and its alloys with high strength and good toughness was also expected by scientists and engineers, which will further broaden their industrial applications.

As studied from the literatures in the past several years, designing the dissimilar-metal blocks with multilayered soft–hard structures can be effectively realized by high pressure torsion + rolling + annealing [10,11], accumulative roll bonding [12,13] and hot pressing + hot rolling + annealing [14]. Multilayered soft–hard structures usually exhibited a good combination of strength and toughness. [15–17]. The enhanced strength was attributed to hetero-deformation induced hardening caused by extra geometrically necessary dislocations accumulated around deformed soft–hard interfaces [10–13,16,18]. The superior impact toughness was explained by that lots of crack deflection or interfacial delamination was formed by high-speed impacting load. These crack deflection or interfacial delamination usually consumed huge energy and impeded crack propagation [17,19]. Recently, Ma et al. [10,12] and Li et al. [15–17] have successfully fabricated novel multilayered soft–hard copper–brass blocks with high yield strength as well as superior impact toughness. Many efforts were devoted to explore the influence factors of strengthening mechanisms for soft–hard copper–brass blocks, such as layer thickness, interfacial bonding strength and hardness ration of soft–hard layer [10–13,16]. However, there was scare exploration of the fracture behavior and toughening mechanism for soft–hard copper–brass blocks. Our recent work indicated that the structural orientation had an important influence on Charpy impact toughness of soft–hard copper–brass block [17]. The Charpy impact toughness tested along the vertical orientation was superior to that test along the parallel orientation. This enhanced impact resistance along the vertical orientation was related to the coordination deformation behavior around copper–brass interfaces [17]. As a matter of fact, many factors could affect the impact toughness for soft–hard copper–brass blocks. Service temperature is one of typical factors that should be paid attention. As reported by previous works [20,21], the face-centered-cubic (FCC) metals usually exhibited slightly variation of impact toughness as the decreasing of testing temperature. This was because of that the impact resistance of FCC metals was usually insensitive to the environmental temperature. For multilayered soft–hard copper–brass blocks, there was a typical feature of many soft–hard interfaces. Up to now, the impact behavior of soft–hard interface under low and high temperatures was not revealed for copper–brass blocks. Impact toughness of the copper–brass blocks was believed to be affected by deformation behaviors of soft–hard interfaces, and it should also be paid more attentions.

In present work, a multilayered soft–hard copper–brass block with high yield strength of ~320 MPa and good uniform elongation of ~20% was prepared by a combined processing technique of diffusion welding, forging and annealing (DWFA technique), which had induced in previous work [17]. The influence of testing temperatures on its impact property was studied. Simultaneously, the fracture mechanisms were revealed at liquid nitrogen temperature (LNT), room temperature (RT) and 200 °C.

2. Experimental

2.1. Materials and Preparation

Commercial ASTM-C11000 copper sheets (99.9 wt% Cu) and ASTM-C26000 brass (Cu-30 wt% Zn) sheets were used in the current work, which were provided by Anhui Xinke New Materials Stock Co., Ltd. The original thicknesses for the copper sheet and brass sheet were 1 mm and 0.8 mm, respectively. Before fabricating a multilayered soft–hard copper–brass block, all of the above sheets were cut with the same dimensions of 100×150 mm^2, and then they were polished using SiC paper ($\Phi 10$ μm for the SiC grit) and washed in an acetone solution for 15 min. Figure 1 shows that a combined DWFA technique was used to prepare the multilayered soft–hard copper–brass blocks. There were four steps in the fabrication processes: (I) 20 layers of copper sheets and 20 layers of brass sheets were stacked at intervals; (II) the stacked copper–brass sheets totaling 40 layers were welded using a diffusion welding machine (ZM-Y, Shanghai Chenhua Electric Furnace Co., Ltd., Shanghai, China) to obtain a copper–brass block, where during the diffusion welding process the stacked copper–brass sheets were extruded under the static pressure of 2 MPa and annealed at 920 °C for 2 h, so as to guarantee the diffusion of Cu/Zn and achieve the metallurgical bonding between the copper and brass sheets; (III) the diffusion-welded copper–brass block was further punched using a pneumatic hammer (C41-75, Nantong Shenwei Forming Machine Works Co., Ltd., Nantong, China), which reduced the thickness from 36 mm to 4 mm and achieved an average layer thickness of ~100 μm; (IV) the forged copper–brass block was finally annealed at 300 for 2.5 h using a muffle furnace (KSL-1100X, HF-Kejing, Hefei, China). The temperature in the chamber was detected using a K-type thermocouple and the measurement accuracy was ±1 °C. The detailed fabricating processes and related parameters can be also found in Figure 1 and in previous work [17].

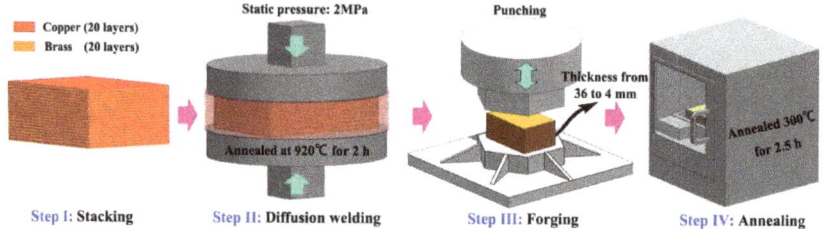

Figure 1. A schematic illustration of the fabrication of a multilayered soft–hard copper–brass block. It is noted that the related processing parameters are included in this figure.

2.2. Mechanical Tests

The microhardness of the copper–brass block was detected utilizing a Vickers hardness tester (HMV-G 21DT, Shimadzu, Tokyo, Japan). The applied load was 0.98 N, and the holding time was 15 s. Each of the hardness values was obtained by averaging at least 5 indents. The uniaxial tension tests were performed at RT using a universal tension machine (LFM-20, Walter+Bai AG, Löhningen, Switzerland). A typical engineering stress–strain curve of the copper–brass block was captured based on a flat "dog bone" tension sample. Its gauge dimensions were $5 \times 2.5 \times 2$ mm^3. During the tension experiment, three copper–brass tension specimens were tested to guarantee the reliability of the tensile result at RT, and the strain rate was 2×10^{-3} s^{-1}. A Charpy impact tester (PH50/15J, Walter+Bai AG, Löhningen, Switzerland) with a testing module with a maximum capacity of 15 J was utilized to assess the impact energy values of the copper–brass blocks. The measurement resolution was 0.001 J. V-notched Charpy impacting specimens were cut with the dimensions of $18 \times 4 \times 2$ mm^3 (length $\times$ height $\times$ thickness mm^3). The detailed specimen dimensions and corresponding images are displayed in Figure 2. In this work, each impact test was performed five times to guarantee the reliability of the data. In order to obtain the various testing temperatures, the impacting specimens were immersed in liquid nitrogen or heated at 200 °C in a furnace for a long time period of 10 min to achieve a specific temperature of −196 °C or 200 °C, respectively. Then, they were immediately taken out for impact testing. All of the above operations were completed within 5 s to guarantee the reliability of the testing temperatures. Detailed information regarding the selection of tension and Charpy impact specimens can be found in Figure 3.

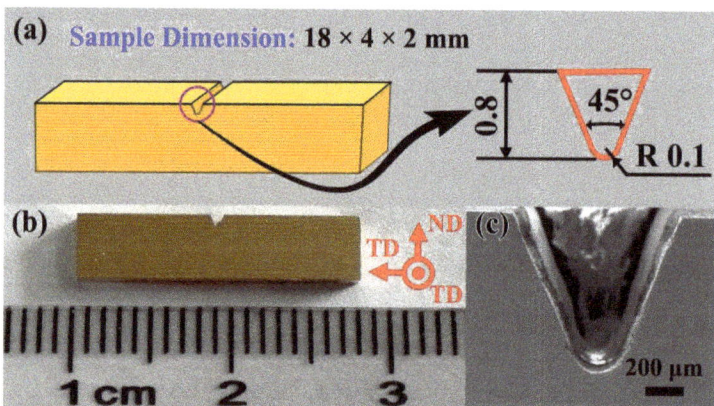

Figure 2. (**a**) The dimensions of the impact specimens in the present work. (**b**) Impact specimens cut from a multilayered soft–hard copper–brass block. (**c**) A scanning electron microscope (SEM) image of the V-notch.

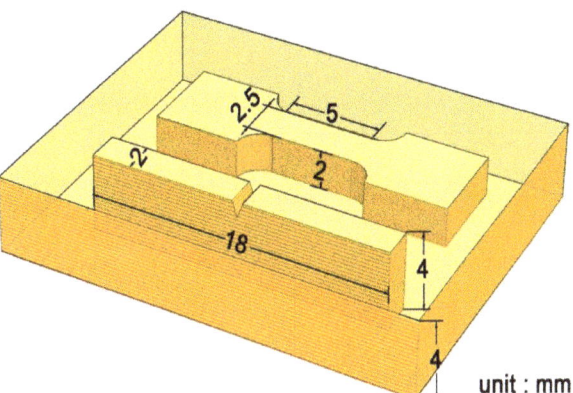

Figure 3. Schematic illustration of the selection of a tension sample and Charpy impact sample.

2.3. Microstructural Characterization

Macro-images of the untested and tested impact samples were obtained using a Nikon camera. A field emission scanning electron microscope (SEM, Quant 250 FEG, FEI, Hillsboro, OR, USA) with an accelerating voltage of 20 keV was used to capture the fracture morphologies for the Charpy impacting samples impacted at various temperatures. The original and deformed microstructures were analyzed using an electron backscattering diffraction (EBSD) technique, which was conducted on the above SEM machine. The accelerating voltage and step size for the EBSD testing were 15 keV and 200–300 nm, respectively. The EBSD specimens were firstly prepared via mechanical polishing, and then they were polished at 3.8 V in a phosphoric acid (85 mL) + H_2O (15 mL) solution for 40 s.

3. Results and Discussion

Figure 4 presents the microstructure of the multilayered copper–brass block. As can be observed from Figure 4a,b, the copper and brass layers exhibit a diverse structure. The copper layer shows a deformed lamellar structure with numerous low-angle boundaries (LAGs), while the brass layer shows an annealed structure consisting of many equiaxed recrystallization grains. The average grain size for the brass layer is ~3.6 μm. The discrepant structures of the copper and brass layers are ascribed to the low recrystallization driving force of the deformed coarse-grained copper, as has been reported in previous studies [16,17]. Figure 4c shows that the hardness value of the copper layer is about 100 HV, which is lower than that of the brass layer (130 HV). This is a typical multilayered soft-hard structure. As reported by Huang et al. [13] and Li et al. [16], the multilayered soft-hard metallic blocks always exhibit a good combination of strength and ductility. As shown in Figure 5a, the presented multilayered soft-hard copper-brass block also shows a high yield strength of ~320 MPa and good uniform elongation of ~20%, which is superior to the mechanical properties of copper-brass layer and the related materials, as already summarized by Huang et al. [13]. In this study, the impact energies of the soft-hard copper-brass block were evaluated under different testing temperatures (LNT, RT, and 200 °C). Figure 5b shows that the highest impact resistance with the impact energy of 8.15 J was achieved at LNT. When the testing temperature increased up to RT, the impact energy was 7.39 J. As the testing temperature increased up to 200 °C, the soft-hard copper-brass block showed a further decrease in impact resistance, and the impact energy reached 7.04 J. The detailed explanation for the relationship between the impact energy and testing temperature can be revealed by further exploring the fracture morphologies and coordinated deformation behavior in the following section.

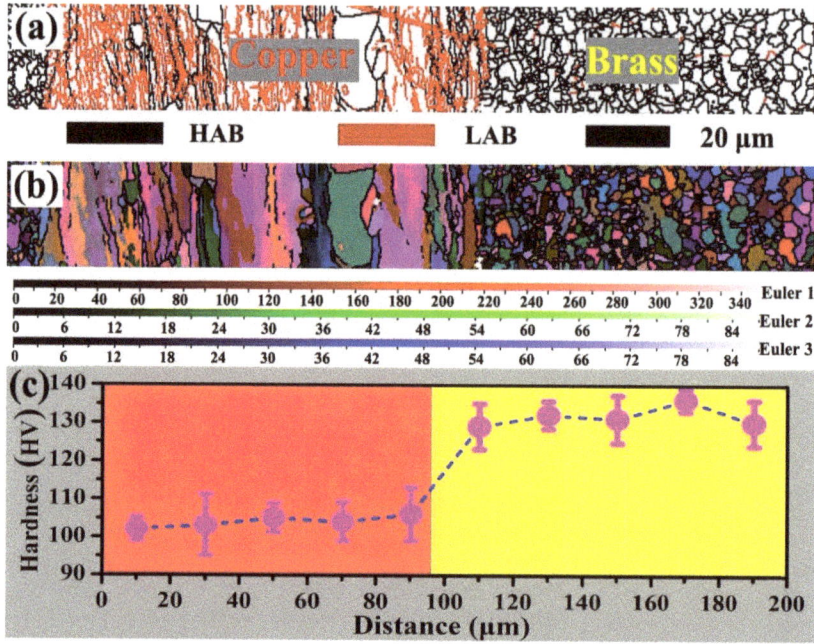

Figure 4. (a) The distributions of the low-angle boundaries (LAGs, misorientation of 2–15°) and high angle-boundaries (HAGs, misorientation of >15°) for the copper–brass block. (b) A Euler map of the cross-sectional copper–brass block. (c) The hardness distribution of the cross-section of copper–brass block.

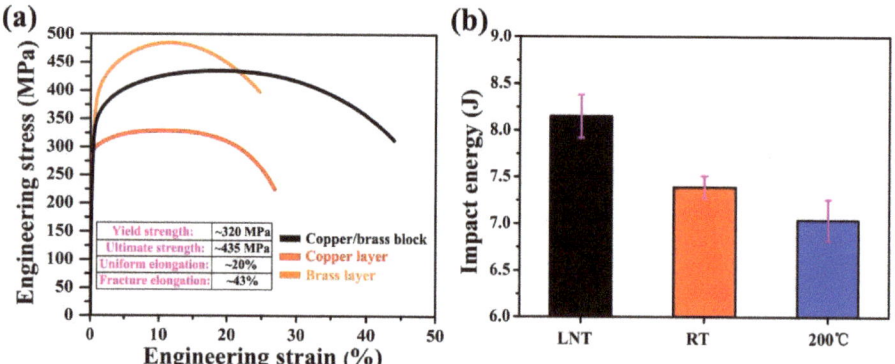

Figure 5. (a) The engineering stress–strain curves of the copper layer, brass layer, and multilayered soft–hard copper–brass block. (b) The impact energies of the present multilayered soft–hard copper–brass block tested at various temperatures.

Figure 6 shows the cross-sectional fracture morphologies around the V-notch for the multilayered soft–hard copper–brass blocks impacted at LNT, RT, and 200 °C. It is noted that all of the impacted samples presented a similar V-crack morphology. The copper and brass layers near the roots of the V-notches were cracked using a high-speed impacting load, while the copper and brass layers far away from the roots of the V-notches suffered from a decreased impacting load and were just bent. In addition, an evident delamination of the copper–brass interface around the root of the V-crack can be found for all impacted samples. Figure 7 clearly indicates that the delamination lengths of copper–brass interface for all impact specimens are nearly identical. As indicated by Osman et al. [19] and Cepeda-Jiménez et al. [22], dissimilar metal blocks with ultrafine laminate structures often show excellent impact toughness, which can be ascribed to the extra delamination of hetero-interfaces, consuming the high plastic deformation energy. The density of the hetero-interfaces will play an important role in determining the impact toughness. Theoretically speaking, a copper–brass block with a thinner layer thickness will have a higher density of hetero-interfaces and will be more likely to achieve a better impact toughness because of the coordinated deformation effects of the numerous hetero-interfaces. In addition, the well-bonded interfaces can accelerate the coordinated deformation effects [16]. A combination of suitable pressure (~0.5–50 MPa) and a high temperature (~0.5–0.8 T_m) can improve the metallurgical bonding between the copper and brass layers, which may enhance the bonding strength for copper–brass interfaces [16] so as to improve the impact toughness. In this work, although the testing temperatures were altered, the cracked brass layers tested at LNT, RT, and 200 °C showed semblable fracture surfaces with many dimples, and the cracked copper layers tested under LNT, RT and 200 °C also show semblable, brittle fracture surfaces (Figure 8a–f). The similar fracture morphologies of the copper–brass layers under various testing temperatures can be ascribed to the impact toughness of metals with FCC crystal lattices, which are usually insensitive to the environmental temperature [20,21]. Thus, there must be other factors that can affect the impact toughness of the present soft–hard copper–brass blocks. As shown in Figure 8g,h, the tiny dimples found on the delaminated surface of copper–brass interface at RT and LNT were denser, which may indicate more energy consumption. Generally speaking, energy consumption around hetero-interfaces is always related to coordinated plastic deformation. For the present copper–brass interfaces, the brass had a nearly dislocation-free structure, which was believed to make a great contribution to the energy consumption. The deformed structures of coarse-grained brass around the delamination copper–brass interfaces are displayed in Figure 9. It shows that the deformation twins decrease with the increase in testing temperature. As confirmed by previous studies [17,23,24], deformation twins are inclined to be formed at high strain rates from the high-speed impact load, especially at low temperatures, and may consume more deformation energy and enhance the impact toughness of the metals. This may be the reason that the impact energy at LNT is higher than that at RT. As compared in Figure 8g–i, the fracture morphology of the delaminated soft–hard interface at 200 °C shows scarce dimples, which may indicate a mild plastic deformation of the copper–brass interfaces. Kulagin et al. [25] and Malik et al. [26] have indicated that the grain boundaries and phase interfaces were weak when a sample was heated at high temperatures. In fact, the grain boundaries and phases are believed to be special structures, which are composed of many defects (vacancies, dislocations, micro-holes, etc.). These special structures usually have a lower softening temperature, which is caused by recovery and recrystallization. Thus, the impact energy of copper–brass at 200 °C has the lowest impact resistance, and the impact energy or toughness is decreased with the increase in testing temperature.

Although this work has systematically explored the effects of three kinds of typical testing temperatures on the impact properties of a copper–brass block with a layer thickness of ~100 μm, some other influences, such as the layer thickness and preparation factors, have been ignored. These will be further revealed in future studies, which may provide more precise theoretical guidance for the industrial applications of copper–brass blocks.

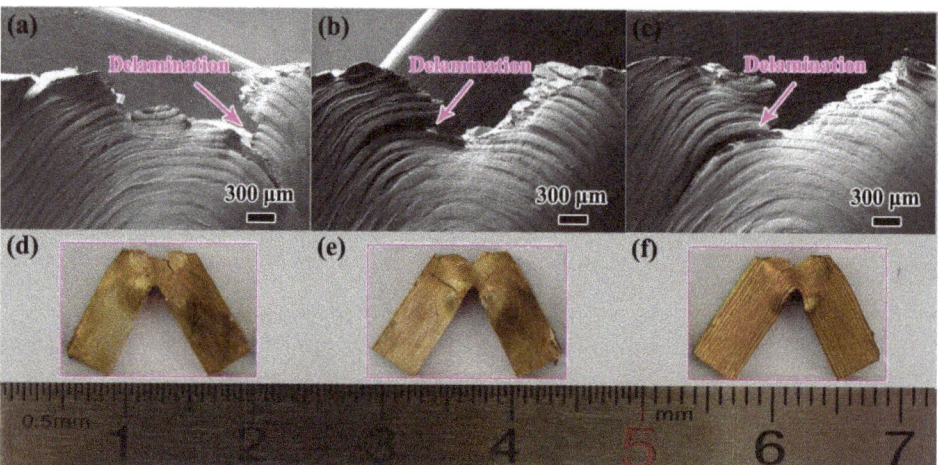

Figure 6. The images of the impacted copper–brass blocks at different temperatures: (**a**–**c**) the SEM images around the V-notch for fractured copper–brass blocks at LNT, RT, and 200 °C, respectively; (**d**–**f**) the optical images for fractured copper–brass blocks at LNT, RT, and 200 °C, respectively.

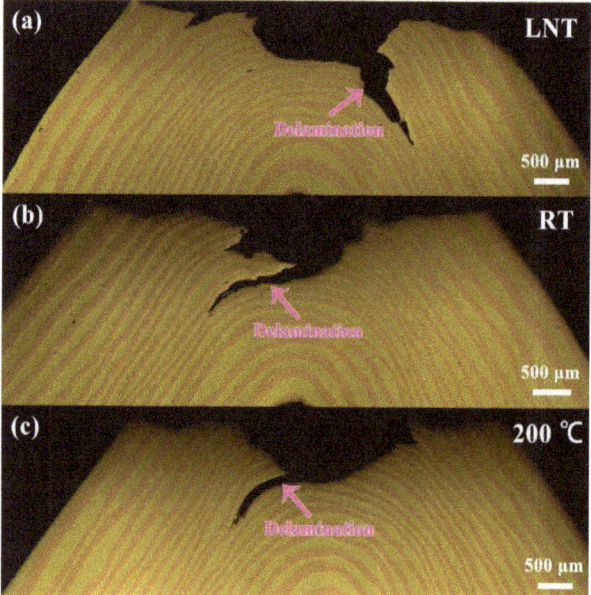

Figure 7. (**a**–**c**) The cross-sectional optical images of the multilayered copper–brass blocks impacted at LNT, RT, and 200 °C, respectively. It is noted that the cross-section is selected in the mid-plane along the thickness direction.

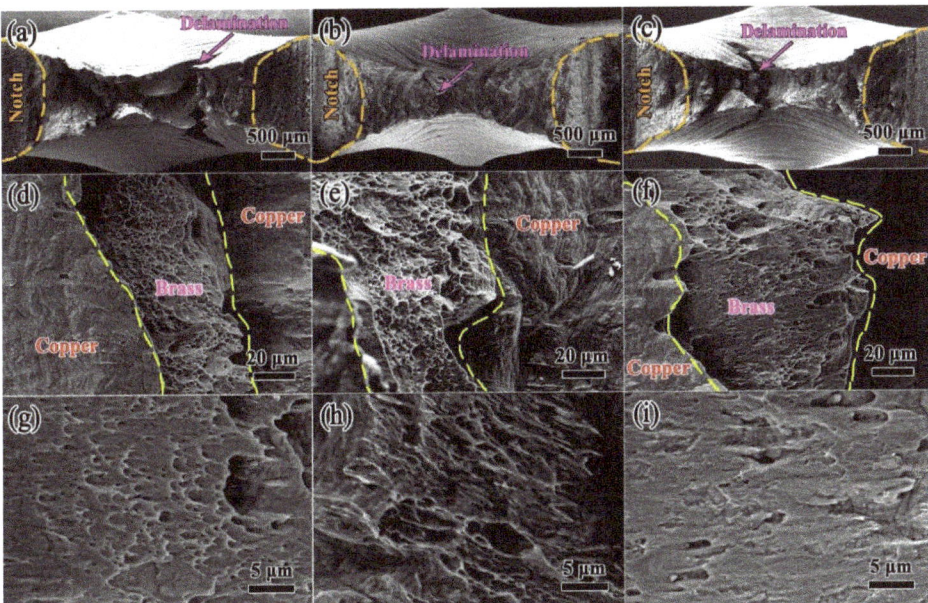

Figure 8. (**a–c**) The fracture surfaces of the copper–brass blocks impacted at LNT, RT, and 200 °C, respectively. (**d–f**) The fracture morphologies of the copper and brass layers at LNT, RT, and 200 °C, respectively. (**g–i**) The morphologies of the delaminated copper–brass interfaces at LNT, RT, and 200 °C, respectively.

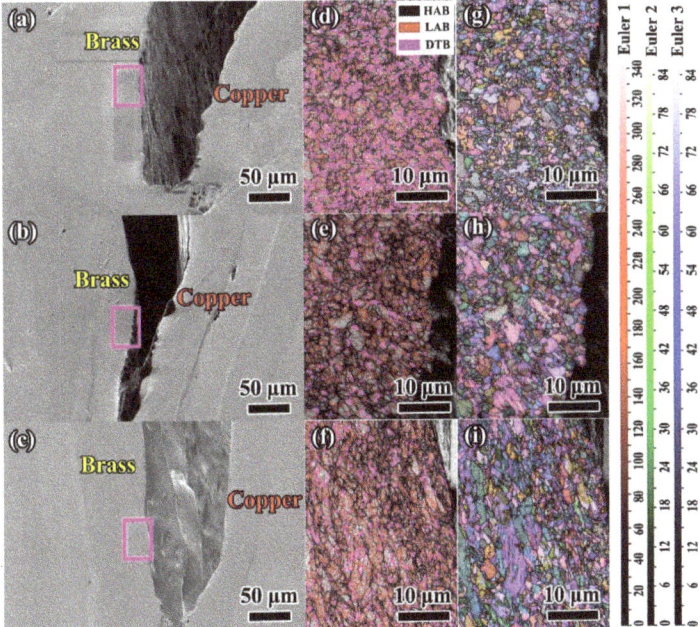

Figure 9. (**a–c**) The SEM images of delaminated copper–brass interfaces at LNT, RT, and 200 °C, respectively. (**d–f**) The distributions of LAGs, HAGs, and deformed twin boundaries (DTBs) of the selected regions in (**a–c**), respectively. (**g–i**) The Euler maps of the selected regions in (**a–c**), respectively.

4. Conclusions

In summary, a combined DWFA technique was employed to successfully prepare a copper–brass block with a soft–hard multilayered structure. The influence of the testing temperature on the impact properties was revealed. Some conclusions were drawn, as follows:

1. The impact energies of the present multilayered soft–hard copper–brass blocks tested at LNT, RT, and 200 °C were 8.15 J, 7.39 J, and 7.04 J, respectively, which indicated that the impact energy was positively dependent on the testing temperature;
2. The copper–brass layers that cracked under various testing temperatures show similar fracture morphologies. This can be ascribed to the fracturing of metals with a FCC crystal lattice usually being insensitive to the environmental temperature;
3. The highest impact energy at LNT was attributed to the high density of tiny dimples caused by coordinated plastic deformation effects during the delamination of the soft–hard copper–brass interfaces. The high temperature of 200 °C can weaken the copper–brass interface, reduce the absorption of the deformation energy, and lead to decreased impact resistance.

Author Contributions: Investigation, T.L., K.L., M.W. and Q.Y.; formal analysis, J.L., Q.M., M.C. and Y.S.; resources, Z.Z. and Y.Z.; methodology, Q.M. and Y.W.; writing—original draft, T.L.; writing—review and editing, Y.Z., Q.M. and J.L. All authors have read and agreed to the published version of the manuscript.

Funding: The authors acknowledge the financial support of the Natural Science Foundation of Anhui Province (2208085QE125), the National Natural Science Foundation of China (52101030), the Scientific Research Starting Foundation of Anhui Polytechnic University (2020YQQ026, S022021005), the Innovation and Entrepreneurship Training Program for College Students of Anhui Province (S202110363167, S202110363174), and the Scientific Research Foundation of Anhui Polytechnic University of China (Xjky2022024).

Institutional Review Board Statement: Not applicable.

Informed Consent Statement: Not applicable.

Data Availability Statement: The data presented in this study are available on request from the corresponding authors.

Conflicts of Interest: The authors declare no conflict of interest.

References

1. Liang, N.N.; Liu, J.Z.; Lin, S.C.; Wang, Y.; Wang, J.T.; Zhao, Y.H.; Zhu, Y.T. A multiscale architectured CuCrZr alloy with high strength, electrical conductivity and thermal stability. *J. Alloys Comp.* **2018**, *735*, 1389–1394. [CrossRef]
2. Lu, X.K.; Zhao, Y.; Wang, G.; Zhu, X.B. Effects of structure characteristics and fluid on the effective thermal conductivity of sintered copper foam. *Results Phys.* **2020**, *19*, 103655. [CrossRef]
3. Mao, Q.Z.; Zhang, Y.S.; Liu, J.Z.; Zhao, Y.H. Breaking material property trade-offs via macrodesign of microstructure. *Nano Lett.* **2021**, *21*, 3191–3197. [CrossRef] [PubMed]
4. Mao, Q.Z.; Zhang, Y.S.; Guo, Y.Z.; Zhao, Y.H. Enhanced electrical conductivity and mechanical properties in thermally stable fine-grained copper wire. *Commun. Mater.* **2021**, *2*, 46. [CrossRef]
5. Liang, N.N.; Zhao, Y.H.; Wang, J.T.; Zhu, Y.T. Effect of grain structure on Charpy impact behavior of copper. *Sci. Rep.* **2017**, *7*, 44738. [CrossRef]
6. An, X.H.; Lin, Q.Y.; Wu, S.D.; Zhang, Z.F. Improved fatigue strengths of nanocrystalline Cu and Cu-Al Alloys. *Mater. Res. Lett.* **2015**, *3*, 135–141. [CrossRef]
7. Shirdel, M.; Mirzadeh, H.; Parsa, M.H. Nano/ultrafine grained austenitic stainless steel through the formation and reversion of deformation-induced martensite: Mechanisms, microstructures, mechanical properties, and TRIP effect. *Mater. Charact.* **2015**, *103*, 150–161. [CrossRef]
8. Straumal, B.B.; Pontikis, V.; Kilmametov, A.R.; Mazilkin, A.A.; Dobatkin, S.V.; Baretzky, B. Competition between precipitation and dissolution in Cu-Ag alloys under high pressure torsion. *Acta Mater.* **2017**, *122*, 60–71. [CrossRef]
9. Tajally, M.; Huda, Z.; Masjuki, H.H. A comparative analysis of tensile and impact-toughness behavior of cold-worked and annealed 7075 aluminum alloy. *Inter. J. Impact Eng.* **2010**, *37*, 425–432. [CrossRef]

10. Ma, X.L.; Huang, C.X.; Xu, W.Z.; Zhou, H.; Wu, X.L.; Zhu, Y.T. Strain hardening and ductility in a coarse-grain/nanostructure laminate material. *Scr. Mater.* **2015**, *103*, 57–60. [CrossRef]
11. Wang, Y.F.; Yang, M.X.; Ma, X.L.; Wang, M.S.; Yin, K.; Huang, A.H.; Huang, C.X. Improved back stress and synergetic strain hardening in coarse-grain/nanostructure laminates. *Mater. Sci. Eng. A* **2018**, *727*, 113–118. [CrossRef]
12. Ma, X.L.; Huang, C.X.; Moering, J.; Ruppert, M.; Hoppel, H.W.; Goken, M.; Narayan, J.; Zhu, Y.T. Mechanical properties of copper/bronze laminates: Role of interfaces. *Acta Mater.* **2016**, *116*, 43–52. [CrossRef]
13. Huang, C.X.; Wang, Y.F.; Ma, X.L.; Yin, S.H.; Hoppel, W.; Goken, M.; Wu, X.L.; Gao, H.J.; Zhu, Y.T. Interface affected zone for optimal strength and ductility in heterogeneous laminate. *Mater. Today* **2018**, *21*, 713–719. [CrossRef]
14. Huang, M.; Xu, C.; Fan, G.H.; Maawad, E.; Gan, W.M.; Geng, L.; Lin, F.X.; Tang, G.Z.; Wu, H.; Du, Y.; et al. Role of layered structure in ductility improvement of layered Ti-Al metal composite. *Acta Mater.* **2018**, *153*, 235–249. [CrossRef]
15. Qin, W.B.; Mao, Q.Z.; Kang, J.J.; Liu, Y.Y.; Shu, D.F.; She, D.S.; Liu, Y.F.; Li, J.S. Superior impact property and fracture mechanism of a multilayered copper/bronze laminate. *Mater. Lett.* **2019**, *250*, 60–63. [CrossRef]
16. Li, J.S.; Wang, S.Z.; Mao, Q.Z.; Huang, Z.W.; Li, Y.S. Soft/hard copper/bronze laminates with superior mechanical properties. *Mater. Sci. Eng. A* **2019**, *756*, 213–218. [CrossRef]
17. Liu, T.; Gu, C.Y.; Li, J.S.; Zhou, Z.C.; Lu, Y.; Gao, F.; Chen, M.; Mao, Q.Z.; Lu, X.K.; Li, Y.S. Effect of structural orientation on the impact properties of a soft/hard copper/brass laminate. *Vacuum* **2021**, *191*, 110388. [CrossRef]
18. Zhu, Y.T.; Wu, X.L. Perspective on hetero-deformation induced (HDI) hardening and back stress. *Mater. Res. Lett.* **2019**, *7*, 393–398. [CrossRef]
19. Osman, T.M.; Hassan, H.A.; Lewandowski, J.J. Interface effects on the quasi-static and impact toughness of discontinuously reinforced aluminum laminates. *Metall. Mater. Trans. A* **2008**, *39A*, 1993–2006. [CrossRef]
20. Ibrahim, O.H.; Ibrahim, I.S.; Khalifa, T.A.F. Impact behavior of different stainless steel weldments at low temperatures. *Eng. Fail. Anal.* **2010**, *17*, 1069–1076. [CrossRef]
21. Smirnov, I.; Konstantinov, A. Influence of ultrafine-grained structure produced by equal-channel angular pressing on the dynamic response of pure copper. *Procedia Struct. Integrity* **2018**, *13*, 1336–1341. [CrossRef]
22. Cepeda-Jiménez, C.M.; García-Infanta, J.M.; Pozuelo, M.; Ruano, O.A.; Carreño, F. Impact toughness improvement of high-strength aluminum alloy by intrinsic and extrinsic fracture mechanisms via hot rolling bonding. *Scr. Mater.* **2009**, *61*, 407–410. [CrossRef]
23. Gludovatz, B.; Hohenwarter, A.; Thurston, K.V.S.; Bei, H.; Wu, Z.; George, E.P.; Ritchie, R.O. Exceptional damage-tolerance of a medium-entropy alloy CrCoNi at cryogenic temperatures. *Nat. Commun.* **2015**, *7*, 10602. [CrossRef] [PubMed]
24. Hasan, M.N.; Liu, Y.F.; An, X.H.; Gu, J.; Song, M.; Cao, Y.; Li, Y.S.; Zhu, Y.T.; Liao, X.Z. Simultaneously enhancing strength and ductility of a high-entropy alloy via gradient hierarchical microstructures. *Inter. J. Plast.* **2019**, *123*, 178–195. [CrossRef]
25. Kulagin, R.; Beygelzimer, Y.; Ivanisenko, Y.; Mazilkin, A.; Straumal, B.; Hahn, H. Instabilities of interfaces between dissimilar metals induced by high pressure torsion. *Mater. Lett.* **2018**, *222*, 172–175. [CrossRef]
26. Malik, A.; Chaudry, U.M.; Hamad, K.; Jun, T.S. Microstructure features and superplasticity of extruded, rolled and SPD-processed magnesium alloys: A short review. *Metals* **2021**, *11*, 1766. [CrossRef]

Article

Evaluation of Biocompatibility of 316 L Stainless Steels Coated with TiN, TiCN, and Ti-DLC Films

Jia Lou [1,2], Beibei Ren [1,2], Jie Zhang [3,*], Hao He [4,*], Zonglong Gao [1] and Wei Xu [5]

1. School of Materials Science and Engineering, Xiangtan University, Xiangtan 411105, China; lou3166@xtu.edu.cn (J.L.); ren3087578860@163.com (B.R.); gaozonglong1996@163.com (Z.G.)
2. Research and Development Department, Taoyuan Institute of Advanced Manufacturing, Foshan 528000, China
3. Department of Orthodontics, Changsha Stomatological Hospital, Changsha 410000, China
4. School of Microelectronics and Materials Engineering, Guangxi University of Science and Technology, Liuzhou 545006, China
5. The Key Lab of Guangdong for Modern Surface Engineering Technology, National Engineering Laboratory for Modern Materials Surface Engineering Technology, Institute of New Materials, Guangdong Academy of Sciences, Guangzhou 510650, China; xuw0907@163.com
* Correspondence: kqzj@hnucm.edu.cn (J.Z.); 100001865@gxust.edu.cn (H.H.)

Abstract: In this study, TiN, TiCN, and Ti-diamond-like carbon (Ti-DLC) films were coated on 316 L stainless steel (AISI 316 L) substrate surface by physical vapor deposition. The biocompatibility of the three films (TiN, TiCN, and Ti-DLC) and three metals (AISI 316 L, Ti, and Cu) was compared on the basis of the differences in the surface morphology, water contact angle measurements, CCK-8 experiment results, and flow cytometry test findings. The biocompatibility of the TiN and TiCN films is similar to that of AISI 316 L, which has good biocompatibility. However, the biocompatibility of the Ti-DLC films is relatively poor, which is mainly due to the inferior hydrophobicity and large amount of sp^2 phases. The presence of TiC nanoclusters on the surface of the Ti-DLC film aggravates the inferior biocompatibility. Compared to the positive Cu control group, the Ti-DLC film had a higher cell proliferation rate and lower cell apoptosis rate. Although the Ti-DLC film inhibited cell survival to a certain extent, it did not show obvious cytotoxicity. TiN and TiCN displayed excellent performance in promoting cell proliferation and reducing cytotoxicity; thus, TiN and TiCN can be considered good orthodontic materials, whereas Ti-DLC films require further improvement.

Keywords: Ti-DLC; biocompatibility; wettability; proliferation; apoptosis

Citation: Lou, J.; Ren, B.; Zhang, J.; He, H.; Gao, Z.; Xu, W. Evaluation of Biocompatibility of 316 L Stainless Steels Coated with TiN, TiCN, and Ti-DLC Films. *Coatings* 2022, 12, 1073. https://doi.org/10.3390/coatings12081073

Academic Editor: Alenka Vesel

Received: 4 July 2022
Accepted: 26 July 2022
Published: 29 July 2022

Publisher's Note: MDPI stays neutral with regard to jurisdictional claims in published maps and institutional affiliations.

Copyright: © 2022 by the authors. Licensee MDPI, Basel, Switzerland. This article is an open access article distributed under the terms and conditions of the Creative Commons Attribution (CC BY) license (https://creativecommons.org/licenses/by/4.0/).

1. Introduction

AISI 316 L is widely used in medical devices such as orthodontic fixed brackets, because of its excellent machining performance, mechanical properties, corrosion resistance, and cost effectiveness [1,2]. However, when exposed to human oral saliva, AISI 316 L undergoes wear and corrosion, such as crevice corrosion, intergranular corrosion, pitting corrosion, and fretting corrosion [3]. When in contact with human bodily fluids over a long period of time, corroded stainless steel (SS) releases chromium or nickel ions, which induces inflammation and cytotoxic effects [4,5]. At the same time, exposed AISI 316 L releases high concentrations of molybdenum ions, which can pose a threat to human health. Moreover, a larger irregular surface area of exposed SS allows a greater capacity for bacterial adsorption, resulting in poor biocompatibility [6].

To overcome these drawbacks, various coating materials have been applied on the SS surface to reduce application defects and enhance biocompatibility [7]. Such coating materials include TiN, TiCN, and Ti-DLC, which can be applied by physical vapor deposition (PVD), chemical vapor deposition (CVD), and other methods. Due to their simple coating procedure, excellent corrosion resistance, high hardness, and good biocompatibility, TiN films have been widely used in various medical fields such as human implants and surgical instruments, especially as orthodontic materials [8]. This film not only improves wear resistance and reduces the corrosion rates of the SS substrate, but also inhibits the release of harmful ions. Thus, the biocompatibility of the SS substrate is significantly improved subsequently, enhances the serviceability of the orthodontic part. Subramanian et al. [9] reported that the excellent biocompatibility of TiN/VN multilayer coatings is due to minimal bacterial adhesion on the film surface. Braic et al. [10] analyzed the biocompatibility of TiN and TiN/TiAlN coatings and found that both coatings showed superior biocompatibility to uncoated SS substrates, in terms of cell density, cell viability, and cell morphology. Some researchers have focused on TiCN films, which exhibit higher properties than TiN films; however, their preparation is much more difficult [11]. Ertuerk et al. [12,13] pointed out that as a solid solution of TiN and TiC, TiCN combines the advantages of both materials and exhibits higher hardness, better tribological properties, and better lubrication than TiN in practical applications. In addition, its non-cytotoxic properties combined with its mechanical and corrosive properties make TiCN a very effective material for biomedical applications. Madaoui et al. [14] demonstrated that XC48 steel plated with TiCN in a 3.5% NaCl solution provided better resistance to uniform and pitting corrosion than bare steel, and effectively prevents the release of matrix ions. Antunes et al. [15] reported that compared to uncoated AISI 316 L, TiCN-coated parts exhibited no cytotoxicity or genotoxicity.

Furthermore, DLC films have proven to be potential biomedical materials with excellent performance, due to the fact that DLC has a higher hardness, better wear resistance, and better chemical inertness than TiN and TiCN [16,17]. To overcome the low bonding strength and high internal stress, DLC films are doped with different metals (Cr, Ti, Zr) to reduce the difference in the thermal expansion coefficients between the DLC films and substrates [18,19]. A Ti-DLC film on SS has smaller internal stress, larger film base bonding, and better mechanical properties such as hardness and toughness. In our previous study, Ti-DLC had the lowest surface roughness, while exhibiting high hardness and low COF [20]. Moreover, the Ti-DLC coating exhibits high corrosion resistance because the TiC crystals block the path of corrosive substances through the film [11]. However, a previous study showed that DLC provides a weak cell adhesion matrix when tested with human mesenchymal stem cells, osteoblasts and osteosarcoma cell lines, proving that the compatibility of Ti-DLC films is debatable [21]. Thus, the application of Ti-DLC films on orthodontically fixed brackets still faces significant challenges.

To overcome this limitation, in this study, TiN, TiCN, and Ti-DLC films were coated on the surface of AISI 316 L by using multi-arc ion plating method [11]. Ti, a commonly used dental implant, and Cu, a powerful antimicrobial metal, were used in the experiment for comparison. This study aimed to select a material for coating on AISI 316 L metal implants that would exhibit excellent in-service properties while improving the biocompatibility of AISI 316 L.

2. Experimental

In this experiment, AISI 316 L (composition given in Table 1) was used as the substrate material, which was cut into sheets with dimensions of 15 mm × 15 mm × 12 mm. Before depositing, the flaky substrate was sanded with #80-#2000 silicon carbide sandpaper. The sheet substrate was polished to a defect-free mirror finish and cleaned ultrasonically with ethanol for 10 min and deionized water for 5 min to remove surface contamination. The as-treated sheet substrates were placed into an arc coating equipment (Damp AS700, ProChina Limited, Beijing, China) to deposit TiN, TiCN, and Ti-DLC films; the three film deposition parameters were derived from a previous study [20]. The surface morphologies of the deposited films were observed through field emission scanning electron microscopy (SEM, FEI, Nova NanoSEM230, FEI Company, Hillsboro, OR, USA). The contact angle of water on the surface of each sample was measured using a contact angle test system (Orbital surface tension meter, Ramé-hart Model 250, Ramé-hart Instrument Company, Succasunna, NJ, USA) at room temperature (27 °C). The specific test technique used was the solid drop method with a drop volume of 2 µL.

Table 1. Composition of AISI 316 L substrate.

Alloy	Main Alloying Elements (wt%)						
-	Cr	Ni	Mo	N	C	Mn	Fe
AISI 316 L	16.30	14.20	1.3	0.06	0.05	2.03	balance

For this experiment, L929 mouse fibroblasts obtained from the Institute of Advanced Study of Central South University (Changsha, China) were used to explore the biocompatibility of each sample. For cell recovery, frozen tubes containing L929 mouse fibroblasts were placed in a 37 °C water bath and thawed by constant shaking. After an initial observation of L929 mouse fibroblast viability through pressed microscopy, the cells were cultured in modified Eagle's medium (DMEM, Gibco, Life Technologies Corporation, Grand Island, NY, USA) containing 10% fetal bovine serum (FBS). After reaching 80%–90% confluence, the adherent cells were washed, trypsinized, counted, and re-suspended to seed on the samples. Each sample was separately inoculated into a 6-well tissue culture plate. The cellular concentration was 1×10^4 mL^{-1} in the medium (DMEM with 10% FBS) and the sample was incubated for 1 week. The culture parameters were as follows: (37 ± 1) °C temperature and 5% CO_2 concentration. The samples were evaluated at 24, 48, and 72 h. The samples containing cells at 24, 48, and 72 h were washed with phosphate-buffered saline (PBS). After cell fixation, dehydration, and drying, the cell adhesion pattern of each sample was observed by SEM at three incubation time points.

All sets of samples from the three time points removed from the culture plate were digested using trypsin and incubated in a 6-well plate for 4 h after adding a CCK-8 reaction solution at 10 mL/well. The absorbance values at 450 nm of the corresponding seven groups of the samples after 24, 48, and 72 h of incubation were measured using an enzyme standardization instrument (ST-360, Dan Ding Shanghai International Trade Co., Ltd., Shanghai, China). Cell proliferation was detected using Cell Counting Kit 8 (CCK-8, Shanghai Beyotime Institute of Biotechnology, Shanghai, China).

Flow cytometry was conducted to detect the rate of apoptosis after staining with Annexin V-FITC/PI Apoptosis Double Staining Kit (bioworlde, BD0062-3, Bioworld Technology, Nanjing, China). All samples were digested using trypsin. Cell suspensions were prepared after washing with PBS and placed in flow-through tubes. Annexin V-FITC/PI was then added, and the samples were incubated at 25 °C for 15 min while being protected from light. The apoptosis rate was calculated on the basis of detection by flow cytometry, which was taken as the sum of the percentages of Q2 (late apoptotic cells) and Q3 (early apoptotic cells).

Tests related to biocompatibility must be repeated at least three times to ensure reproducibility. The data obtained from the experiments were statistically analyzed using a relevant software, and the measures conforming to the normal distribution were expressed as x ± s. One-way ANOVA was used to compare means between groups, and the least significant difference method was used for two-way comparison. The test level was set at $p < 0.05$. Biocompatibility evaluation criteria for relevant in vitro cytotoxicity tests are in accordance with ISO 10993-5.

3. Results and Discussions

3.1. Surface Characteristics

The surface morphologies of the films are shown in Figure 1. The surface of the SS substrate was basically smooth and flat, but there were some tiny pores with irregular shapes, which was mainly due to the cutting and polishing effect. After deposition, few pores were observed on the surfaces of the TiN and TiCN films, while some micro-sized TiC particles were observed. The Ti-DLC film exhibited the lowest porosity, but several flaky defects and clusters were observed on its surface. The adhesion of the three films to the substrate is firm [20]. The thicknesses of the TiN, TiCN, and Ti-DLC films were 2.05, 4.10, and 4.48 µm, respectively; further information about the film layer characteristics and properties can be obtained from previous studies [20]. The SEM images depicting the morphology of the TiC particles of TiCN and Ti-DLC are shown in Figure 2. The TiC particles in the TiCN film were nearly spherical, and their diameters were 1–2 µm; however, TiC clusters on the Ti-DLC films exhibited irregular shapes. Their diameters were in the sub-micrometer or nanometer range. The number of TiC clusters on the Ti-DLC films was significantly higher than that on the TiCN films.

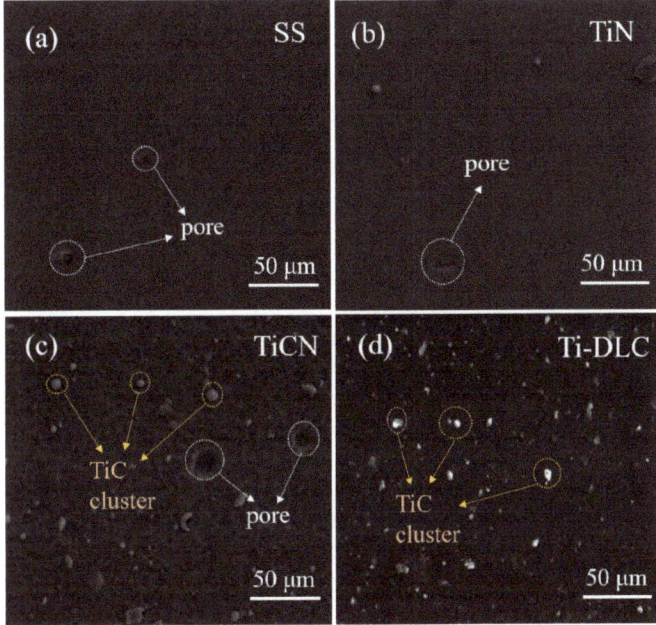

Figure 1. SEM images of the surfaces of (a) SS; (b) TiN; (c) TiCN; (d) Ti-DLC.

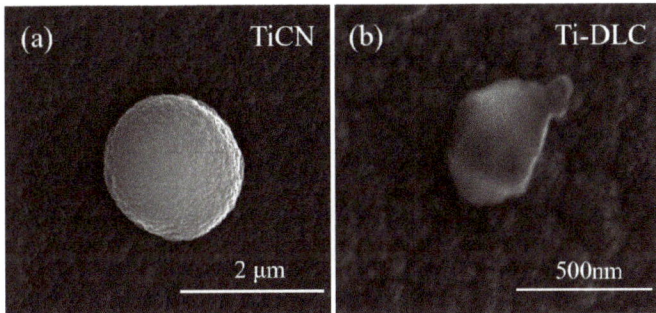

Figure 2. SEM images showing the morphology of TiC particles in the films of (a) TiCN and (b) Ti-DLC.

According to our previous results on XRD [20], peaks of C and TiC were observed in both the TiCN and Ti-DLC films, but the latter was more pronounced, indicating that more TiC was formed. From our previous experimental results of Raman spectroscopy [20], the response intensity for sp^2-hybridized C in Ti-DLC films was approximately 50 times higher than that for sp^3-hybridized C. It was found that a certain amount of sp^3-hybridized C was present in the Ti-DLC films, and the D peak area was significantly larger than the G peak area (I_D/I_G = 2.79). This indicates that the proportion of sp^2-hybridized C was higher than that of sp^3-hybridized C.

Figure 3 shows the water contact angle images of all the samples. The water contact angle strongly affects cell adhesion and cell activity [22]. As shown in Figure 3a, SS is hydrophobic, which may be related to its surface porosity and roughness. The TiN and TiCN films substantially improved the wettability of the AISI 316 L substrates. Both films exhibited strong hydrophilicity. Similar results were reported by Khan et al. for TiN [23] and Sunthornpan et al. for TiCN [24]. When AISI 316 L was coated with a Ti-DLC film, the water contact angle increased to 96.1° ± 0.4° and showed significant hydrophobicity, which is mainly due to the chemical properties of DLC. Graphite and other sp^2 carbon materials are recognized as hydrophobic materials with a water contact angle of ~90°, even when the surface is smooth [25,26]. Some researchers have reported that the surfaces of these materials are susceptible to the adsorption of hydrocarbon contaminants from the air environment [27]. The water contact angle of sp^3-dominated diamond was slightly lower than that of graphite [28]; however, the proportion of sp^2-hybridized C in the Ti-DLC films was found to be higher than that of sp^3-hybridized C. More importantly, the water contact angle of the Ti-DLC film was greater than 90°. This phenomenon can be attributed to the following three factors. (1) The hydrocarbon contaminants in the air cause Ti-DLC to have a larger water contact angle [29]. (2) As Ti-DLC films contain more sp^2 phases and sp^2-hybridized C have sharp structures on the surface. These sharp structures are arranged in a jagged pattern to give the film layer a large water contact angle. (3) The presence of a large number of TiC nanoclusters with irregular shapes and rough surfaces increased the surface roughness of the Ti-DLC film and thus a large water contact angle.

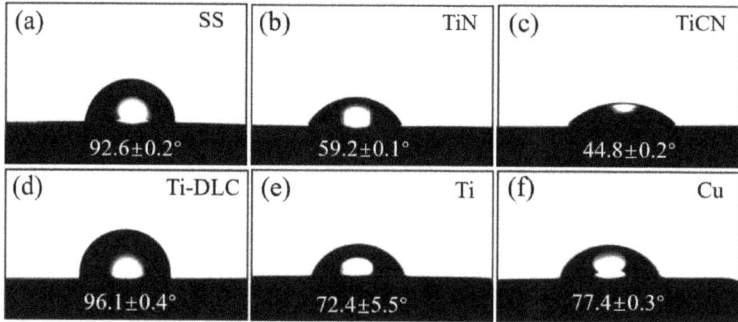

Figure 3. Water contact angle of (**a**) SS; (**b**) TiN; (**c**) TiCN; (**d**) Ti-DLC; (**e**) Ti; (**f**) Cu.

3.2. Biocompatibility Assay

3.2.1. Cell Adhesion Morphology

Figure 4 shows the morphology of L929 mouse fibroblasts after 24, 48, and 72 h of culture for each sample group. As can be seen from the graph, as the incubation time increased, the blank group exhibited better cell growth morphology than the others. The cells were mainly shuttle-shaped and spread well on the surface of the culture plate. For the 24–72 h culture time, it was found that the number of cells kept on increasing, and the cell spreading status improved with increasing culture time. The L929 cells grew well on the surface of the TiCN and TiN films and SS substrates, among which the cells on TiCN spread maximally on the surfaces of the culture plates. Cell adhesion for the Ti group was slightly weaker than that for the other three film samples. The cells on the Ti-DLC surface had the worst spreading status among those on the three films; this result was similar to that observed for the positive Cu control group. The difficulty in cell adhesion on the Ti-DLC surface is due to its hydrophobicity. More importantly, the Ti-DLC film contains more sp^2 phases, and sp^2-hybridized C typically has sharper structures on its surface. These sharp structures directly affect the forces between the polar groups of the cellular proteins and the surface of the material, resulting in weak cell adhesion patterns on the Ti-DLC surface. Although it has rarely been proven that a higher sp^2 ratio negatively affects the cell adhesion of DLC-type coatings, nanodiamond (NCD) and other carbon-based films exhibit similar results. Wang et al. [30] deposited micro-diamond (MCD) and NCD on the surface of a TC4 alloy and found that a higher content of the sp^2 phase in NCD resulted in a lower cell adhesion morphology and reduced cell activity in L929.

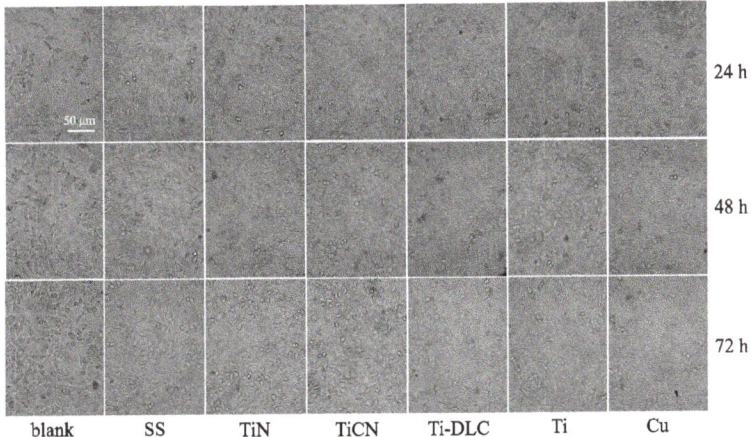

Figure 4. Morphology of L929 cells for the samples after 24, 48, and 72 h of incubation.

3.2.2. CCK-8 Results

Figure 5 shows the absorbance OD values of the culture solution measured at 450 nm for each group at 24, 48, and 72 h of incubation. As shown in Figure 5, the cell counts measured in the SS and TiN groups were similar to those in the blank group for the three studied culture times. The TiN group had the highest number of viable cells and exhibited better biocompatibility than the other groups. The TiCN group had a slightly higher number of cells than the Ti group, especially at 72 h. This indicates that the growth of the cells on the surfaces of SS, TiN, and TiCN was not significantly inhibited and better cell proliferation was demonstrated. The number of cells on the Ti-DLC surface was significantly lower than that on the Ti surface, but much greater than that on the Cu surface.

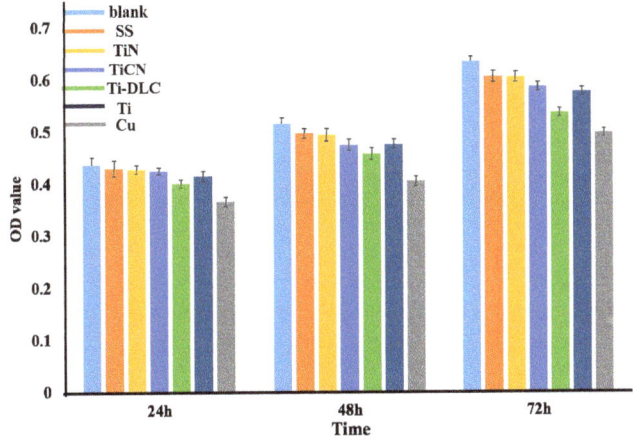

Figure 5. Cell absorbance OD value for the samples after 24, 48, and 72 h of incubation.

The cell proliferation rate is one of the key factors used to determine the biocompatibility of films. Figure 6 shows the relative cell proliferation rate of each sample cultured for 24, 48, and 72 h. Moreover, it is evident from the figure that the cell proliferation rate curves of the SS and TiN groups almost completely overlap. This indicates that there was no difference between the inhibitory effects of SS and TiN on the cells. The cell proliferation rate of the TiCN group was slightly lower than those of the SS and TiN groups. Among the three films, the Ti-DLC group exhibited the lowest cell proliferation rate. The TiN and TiCN groups have higher cell proliferation rates, possibly because they have lower surface roughness and a smaller water contact angle, which is favorable for cell adhesion and growth. The low cell multiplication rate for the Ti-DLC sample was mainly due to its high hydrophobicity. It has been experimentally demonstrated that as the water contact angle increases, the wettability of the material decreases, preventing the cells from attaching to its surface, swelling, and growing well. Eventually, the number of cells, proliferation rates, and differentiation levels are reduced [31].

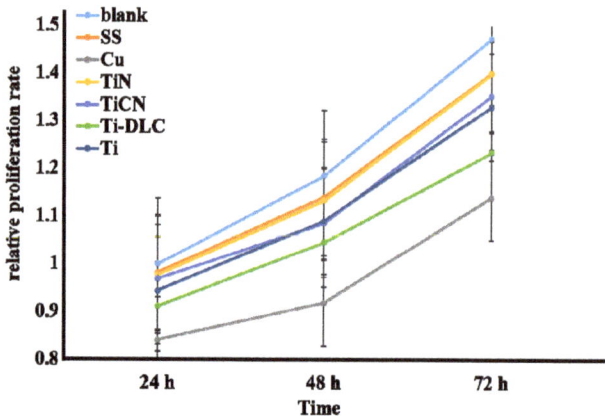

Figure 6. Relative cell proliferation rate for samples after 24, 48, and 72 h of incubation.

3.2.3. Flow Cytometry Results

The results of the flow cytometry tests are presented in Table 2. The apoptosis rate was the sum of the Q2 and Q3 quadrant percentages on a two-dimensional scatter plot. Figures 7 and 8, respectively, show the cell flow diagram and total apoptosis rate of each sample at 24, 48, and 72 h of incubation.

Table 2. Flow cytometry test results.

One-Way ANOVA	Mean Diff			Significant		
	24 h	48 h	72 h	24 h	48 h	72 h
Blank vs. SS	−1.52333	−1.79667	−1.71000	0.156N	0.106Y	0.013Y
Blank vs. Cu	−20.30333	−20.80333	−23.36333	0.000Y	0.000Y	0.000Y
Blank vs. TiN	−1.17000	−1.04333	−1.45667	0.380N	0.560Y	0.014Y
Blank vs. TiCN	−0.89	−1.12	−1.73667	0.666N	0.484Y	0.003Y
Blank vs. Ti-DLC	−10.31	−11.87	−12.13	0.000Y	0.000Y	0.000Y
Blank vs. Ti	−5.98	−6.73333	−7.02333	0.000Y	0.000Y	0.000Y
SS vs. Cu	−18.78000	−19.00667	−21.65333	0.000Y	0.000Y	0.000Y
SS vs. TiN	−0.35333	0.75333	0.25333	0.994N	0.873N	0.989Y
SS vs. TiCN	−0.63333	0.67667	−0.02667	0.897N	0.892N	1.000Y
SS vs. Ti-DLC	−8.78667	−10.07333	−10.42000	0.000Y	0.000Y	0.000Y
SS vs. Ti	−4.45667	−4.93667	−5.31333	0.000Y	0.000Y	0.000Y
Cu vs. TiN	19.13333	19.76000	21.90667	0.000Y	0.000Y	0.000Y

Table 2. Cont.

One-Way ANOVA	Mean Diff			Significant		
	24 h	48 h	72 h	24 h	48 h	72 h
Cu vs. TiCN	19.41333	19.68333	21.62667	0.000Y	0.000Y	0.000Y
Cu vs. Ti-DLC	9.99333	8.93333	11.23333	0.000Y	0.000Y	0.000Y
Cu vs. Ti	14.32333	14.07000	16.34000	0.000Y	0.000Y	0.000Y
TiN vs. TiCN	0.28000	−0.07667	−0.28000	0.998N	1.000N	0.982N
TiN vs. Ti-DLC	−9.14000	−10.82667	−10.67333	0.000Y	0.000Y	0.000Y
TiN vs. Ti	−4.81000	−5.69000	−5.56667	0.000Y	0.000Y	0.000Y
TiCN vs. Ti-DLC	−9.42000	−10.75000	−10.39333	0.000Y	0.000Y	0.000Y
TiCN vs. Ti	−5.09000	−5.61333	−5.28667	0.000Y	0.000Y	0.000Y
Ti-DLC vs. Ti	4.33000	5.13667	5.10667	0.000Y	0.000Y	0.000Y

Note: 24 h: F = 518.300, $p < 0.0001$; 48 h: F = 501.597, $p < 0.0001$; 72 h: F = 2103.81, $p < 0.0001$.

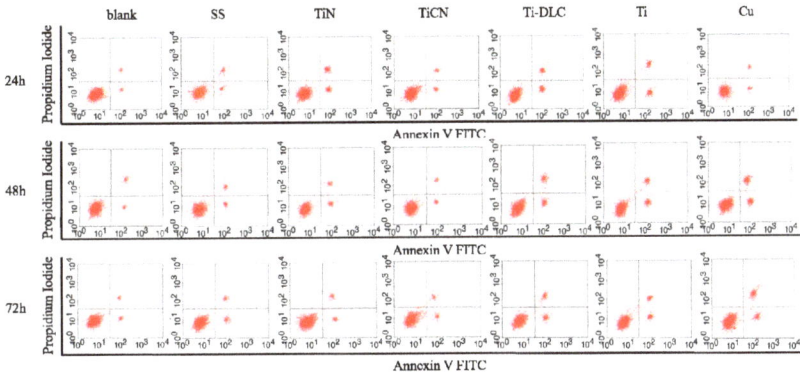

Figure 7. Flow cytometry of samples after 24, 48, and 72 h of incubation.

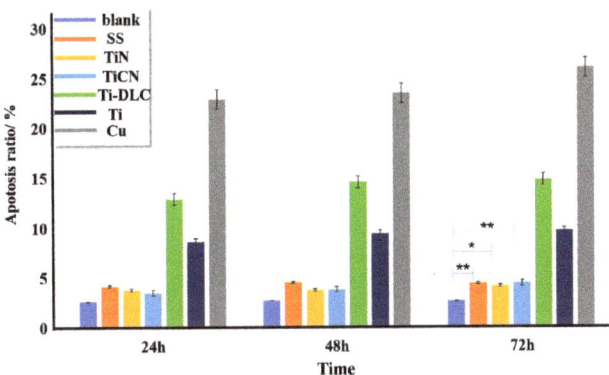

Figure 8. Total apoptosis rate of the samples after 24, 48, and 72 h of incubation.

As can be seen from Figure 8, the blank group had a lower apoptosis rate than the experimental group. The apoptosis rate in the SS group was slightly higher than that in the TiN and TiCN groups but still had a lower apoptosis rate compared to the other experimental groups. This is because the SS surface reacts readily with air to produce a smooth and dense oxide film dominated by oxides of iron and chromium. This film has a certain corrosion resistance, but is still weaker than that of copper and titanium. When placed in harsh environments, such as Cl$^-$ and acidic environments, this film is highly susceptible to pitting corrosion over the long term. Corroded SS releases Cr or Ni ions when present in the oral cavity over time, which in turn causes inflammation and cytotoxic effects [3].

The apoptosis rates of the TiN and TiCN groups were similar at the three time points of 24, 48, and 72 h, and their apoptosis rates were the lowest among the experimental groups. Compared with the apoptosis rate at 24 h, the rate of the TiN film increased by 0.04% at 48 h and 0.69% and 72 h. In contrast, the apoptosis rates of the TiCN films tested at 48 and 72 h increased by 0.32% and 0.63%, respectively. The total apoptosis rate in the SS, TiN, and TiCN groups was less than 5% at all three time points. SS releases harmful ions and thus exhibits a higher rate of apoptosis than TiN and TiCN. TiN has a uniformly dense passivation film and good wettability, which are favorable for cell adhesion and multiplication. As a result, the TiN film exhibited lower apoptosis. Our previous studies have shown that TiC crystals enable TiCN films to exhibit higher corrosion resistance by blocking corrosive substances from penetrating the film path [11]. The high corrosion resistance and hydrophilicity of the TiCN film resulted in the lowest apoptosis rate.

The apoptosis rate in the Ti control group is significantly higher than the three experimental groups mentioned above, and there is a significant difference ($p < 0.01$). Although Ti has a relatively low water contact angle, water contact angle measurements were performed under pure water conditions and short-term action. When Ti is placed in the culture solution for a long time, its passivation film is disrupted and breaks into TiO_2 particles. This leads to a reduction in the hydrophilicity of the Ti surface, and adhesion decreases due to the continuously increasing surface roughness. McGuff et al. [32] studied histological specimens of septic granuloma and peripheral giant cell granuloma in two patients with peri-implant mucosal enlargement in the oral cavity. They found that broken Ti particles caused these reactive lesions. The highest apoptosis rate was observed in the positive Cu control group. In particular, the total apoptosis rate in the positive Cu control group was as high as $(26.09 \pm 0.75)\%$ when the cells were cultured for up to 72 h. Cu releases large amounts of Cu ions when it is placed in the culture solution for a long time. High concentrations of Cu ions can cause cellular inhibition and toxicity in humans [33].

The Ti-DLC group had the highest apoptosis rate among the three films, with a total apoptosis rate in the range of 10%–15% at the three time points. The water contact angle of Ti-DLC was $96.1° \pm 0.4°$, which indicates hydrophobicity and suggests that it is not conducive to cell reproduction and growth. In contrast, the sp^2 phase is more abundant in Ti-DLC, and the sp^2 phases with sharp structures are distributed in a sawtooth shape. By disrupting the cell structure, the Ti-DLC films exhibited more pronounced apoptosis than the other experimental groups [34,35]. Nevertheless, the relevant experimental data show that the cell proliferation rate detected on the surface of the Ti-DLC film at the three culture times was significantly higher than that of the positive Cu control group ($p < 0.05$), while the apoptosis rate was significantly lower than that of the Cu group ($p < 0.01$). This indicates that Ti-DLC does not inhibit cell proliferation or promote apoptosis similar to heavy metals. Moreover, the apoptosis rate of Ti-DLC was only approximately 5% higher than that of Ti at the three culture times. Pure Ti is commonly used for dental implants. Therefore, it is possible to reduce the apoptosis rate of Ti-DLC to that of Ti by improving the preparation process. Thomsonet et al. [36] assessed cytotoxicity by measuring the activity of β-N-acetyl-D-glucosaminidase in a culture medium of primary peritoneal macrophages from DLC-surfaced mice. The results showed no significant difference in enzyme levels between coated and uncoated pores, and no evidence of cell damage to macrophages on the DLC film surface. It can be concluded that Ti-DLC is not cytotoxic, but less biocompatible than TiN and TiCN films. Our previous studies have shown that Ti-DLC films exhibit the most stable electrochemical properties with excellent corrosion resistance in the presence or absence of artificial saliva and at different concentrations of Cl^- and H^+ [11]. Therefore, Ti-DLC films still hold promise for exploration and development in the biomedical field. The biocompatibility of Ti-DLC can be enhanced by improving the preparation method to reduce hydrophobicity and by enhancing the surface properties or reducing the sp^2 phase to allow for a better cell attachment state.

4. Conclusions

The biocompatibility of the TiN, TiCN, and Ti-DLC films deposited on AISI 316 L substrates was compared to determine the most suitable bio-coating material for orthodontic dentistry. The main findings of this study are as follows:

(1) TiN and TiCN had small water contact angles and exhibited significant hydrophilicity. However, Ti-DLC had a relatively large water contact angle and exhibited hydrophobic behavior. The hydrophobicity is mainly due to the large ratio of sp^2 phases. The presence of jagged sp^2 phases and TiC nanoclusters aggravated hydrophobicity.
(2) TiN exhibited the highest cell value-added rate among the three film samples, followed by TiCN. Ti-DLC exhibited the lowest cell proliferation rate due to its high hydrophobicity and sharp sp^2 phase shape.
(3) TiN and TiCN films had lower apoptosis rates than SS because of their excellent corrosion resistance. However, the biocompatibility of Ti-DLC is slightly inferior to that of Ti, but significantly better than that of Cu, mainly because of the hydrophobic nature of Ti-DLC. However, Ti-DLC did not exhibit significant cytotoxicity.
(4) The overall biocompatibility of Ti-DLC was slightly lower than that of the Ti control. Ti-DLC still has potential in the biomedical field, but its preparation must be improved to reduce the amounts of sp^2-hybridized C and TiC nanoclusters and surface roughness.
(5) The presence of TiC nanoclusters with irregular shapes increased the surface roughness of the film with TiC. Thus, a large water contact angle is obtained, which is detrimental to cell adhesion and reproduction.

Author Contributions: Conceptualization, J.L. and J.Z.; methodology, J.Z.; software, B.R.; validation, J.L., B.R. and H.H.; formal analysis, J.Z.; investigation, J.Z.; resources, W.X.; data curation, B.R.; writing—original draft preparation, B.R.; writing—review and editing, J.L., B.R. and Z.G.; visualization, J.Z.; supervision, H.H.; project administration, J.Z.; funding acquisition, J.L., H.H. and W.X. All authors have read and agreed to the published version of the manuscript.

Funding: National Natural Science Foundation of China (52164042). Open project of Foshan Taoyuan Institute of Advanced Manufacturing (TYKF202203004). The Guangdong Academy of Sciences Project of Science and Technology Development (2020GDASYL-20200103109). Guangdong Basic and Applied Basic Research Foundation (2019A1515110710 and 2021A1515012086).

Institutional Review Board Statement: Not applicable.

Informed Consent Statement: Not applicable.

Data Availability Statement: The authors confirm that the data supporting the findings of this study are available within the article.

Conflicts of Interest: The authors declare no conflict of interest.

References

1. Littlewood, S.J.; Millett, D.T.; Doubleday, B.; Bearn, D.R.; Worthington, H.V. Retention procedures for stabilising tooth position after treatment with orthodontic braces. *Cochrane Database Syst. Rev.* **2016**, *1*, CD002283. [CrossRef] [PubMed]
2. Divya, P.; Banswada, S.; Kukunuru, S.; Kavya, K.; Rathod, R.; Polavarapu, K. To compare the accuracy of 0.022 inch slot of stainless steel and ceramic orthodontic brackets marketed by different manufacturers. *J. Pharm. Bioallied Sci.* **2021**, *13*, 1037–1041.
3. Yang, K.; Ren, Y.B. Nickel-free austenitic stainless steels for medical applications. *Sci. Technol. Adv. Mater.* **2010**, *11*, 014105. [CrossRef] [PubMed]
4. Xu, X.; Wang, L.; Wang, G.; Jin, Y. The effect of REDV/TiO$_2$ coating coronary stents on in-stent restenosis and re-endothelialization. *Biomater. Appl.* **2017**, *31*, 911–922. [CrossRef]
5. Kapnisis, K.K.; Pitsillides, C.M.; Prokopi, M.S.; Lapathitis, G.; Karaiskos, C.; Eleftheriou, P.C.; Brott, B.C.; Anderson, P.G.; Lemons, J.E.; Anayiotos, A.S. In vivo monitoring of the inflammatory response in a stented mouse aorta model. *Biomed. Mater. Res. Part A* **2016**, *104*, 227–238. [CrossRef] [PubMed]
6. Kaliaraj, G.S.; Vishwakarma, V.; Kirubaharan, A.M.K. Biocompatible Zirconia-Coated 316 stainless steel with anticorrosive behavior for biomedical application. *Ceram. Int.* **2018**, *44*, 9780–9978. [CrossRef]
7. Grill, A. Diamond-like carbon coatings as biocompatible materials—an overview. *Diam. Relat. Mater.* **2003**, *12*, 166–170. [CrossRef]
8. Jun, Z.; Xie, Y.; Zhang, J.; Wei, Q.; Zhou, B.; Luo, J. TiN coated stainless steel bracket: Tribological, corrosion resistance, biocompatibility and mechanical performance. *Surf. Coat. Technol.* **2015**, *277*, 227–233.

9. Subramanian, B.; Ananthakumar, R.; Kobayashi, A.; Jayachandran, M. Surface modification of 316 L stainless steel with magnetron sputtered TiN/VN nanoscale multilayers for bio implant applications. *Mater. Sci. Mater. Med.* **2012**, *23*, 329–338. [CrossRef] [PubMed]
10. Braic, M.; Balaceanu, M.; Braic, V.; Vladescu, A.; Pavelescu, G.; Albulescu, M. Synthesis and characterization of TiN, TiAlN and TiN/TiAlN biocompatible coatings. *Surf. Coat. Technol.* **2005**, *200*, 1014–1017. [CrossRef]
11. Lou, J.; Gao, Z.L.; Zhang, J.; He, H.; Wang, X.M. Comparative Investigation on Corrosion Resistance of Stainless Steels Coated with Titanium Nitride, Nitrogen Titanium Carbide and Titanium-Diamond-like Carbon Films. *Coatings* **2021**, *11*, 1543. [CrossRef]
12. Ertuerk, E.; Knotek, O.; Burgmer, W.; Prengel, H.G.; Heuvel, H.J.; Dederichs, H.G.; Stossel, C. Ti(CN) coatings using the arc process. *Surf. Coat. Technol.* **1991**, *46*, 39. [CrossRef]
13. Chen, R.; Tu, J.P.; Liu, D.G.; Mai, Y.J.; Gu, C.D. Microstructure, mechanical and tribological properties of TiCN nanocomposite films deposited by DC magnetron sputtering. *Surf. Coat. Technol.* **2011**, *205*, 5228. [CrossRef]
14. Madaoui, N.; Saoula, N.; Zaid, B.; Saidi, D.; Ahmed, A.S. Structural, mechanical and electrochemical comparison of TiN and TiCN coatings on XC48 steel substrates in NaCl 3.5% water solution. *Appl. Surf. Sci.* **2014**, *312*, 134–138. [CrossRef]
15. Antunes, R.A.; Rodas, A.C.D.; Lima, N.B.; Higa, O.Z.; Costa, I. Study of the corrosion resistance and in vitro biocompatibility of PVD TiCN-coated AISI 316 L austenitic stainless steel for orthopedic applications. *Surf. Coat. Technol.* **2010**, *205*, 2074–2081. [CrossRef]
16. Jo, Y.J.; Zhang, T.F.; Son, M.J.; Kim, K.H. Synthesis and electrochemical properties of Ti-doped DLC films by a hybrid PVD/PECVD process. *Appl. Surf. Sci.* **2018**, *433*, 1184–1191. [CrossRef]
17. Qiang, L.; Zhang, B.; Zhou, Y.; Zhang, J. Improving the internal stress and wear resistance of DLC film by low content Ti doping. *Solid State Sci.* **2013**, *20*, 17–22. [CrossRef]
18. Li, W.S.; Zhao, Y.T.; He, D.Q.; Song, Q.; Sun, X.W.; Wang, S.C.; Zhai, H.M.; Zheng, W.W.; Robert, J.K. Optimizing mechanical and tribological properties of DLC/Cr3C2-NiCr duplex coating via tailoring interlayer thickness. *Surf. Coat. Technol.* **2022**, *434*, 128198. [CrossRef]
19. Wang, D.Y.; Chang, Y.Y.; Chang, C.L. Deposition of diamond-like carbon films containing metal elements on biomedical Ti alloys. *Surf. Coat. Technol.* **2005**, *200*, 2175–2180. [CrossRef]
20. Zhang, J.; Lou, J.; He, H.; Xie, Y. Comparative investigation on the tribological performances of TiN, TiCN, and Ti-DLC film-coated stainless steel. *JOM* **2019**, *71*, 4872–4879. [CrossRef]
21. Calzado-Martin, A.; Saldana, L.; Korhonen, H.; Soininen, A.; Kinnari, T.J.; Gomez-Barrena, E.; Tiainen, V.M.; Lappalainen, R.; Munuera, L.; Konttinen, Y.T. Interactions of human bone cells with diamond-like carbon polymer hybrid coatings. *Acta Biomater.* **2010**, *6*, 3325–3338. [CrossRef] [PubMed]
22. Peng, F.; Lin, Y.L.; Zhang, D.D.; Ruan, Q.D.; Tang, K.W.; Li, M.; Liu, X.Y.; Paul, K.C.; Zhang, Y. Corrosion Behavior and Biocompatibility of Diamond-like Carbon-Coated Zinc: An In Vitro Study. *ACS Omega* **2021**, *6*, 9843–9851. [CrossRef] [PubMed]
23. Khan, S.; Chen, S.N.; Ma, Y.C.; Haq, M.U.; Li, Y.D.; Nisar, M.; Khan, R.; Liu, Y.; Wang, J.X.; Han, G.R. Structural and hydrophilic properties of TiN films prepared by ultrasonic atomization assisted spray method under low temperature. *Surf. Coat. Technol.* **2020**, *393*, 12582. [CrossRef]
24. Sunthornpan, N.; Watanabe, S.; Moolsradoo, N. Corrosion resistance and cytotoxicity studies of DLC, TiN and TiCN films coated on 316 L stainless steel. *Siam Physics Congress.* **2018**, *1144*, 012013 (SPC2018). [CrossRef]
25. Adamson, A.W.; Gast, A.P. *The Solid-Liquid Interface- Contact Angle: Experimental Methods and Measurements of Contact Angle*; John Wiley & Sons, Inc.: New York, NY, USA, 1997; pp. 362–372.
26. Raj, R.; Maroo, S.C.; Wang, E.N. Wettability of graphene. *Nano Lett.* **2013**, *13*, 1509–1515. [CrossRef]
27. Kozbial, A.; Zhou, F.; Li, Z.T.; Liu, H.T.; Li, L. Are Graphitic Surfaces Hydrophobic? *Acc. Chem. Res.* **2016**, *49*, 2765–2773. [CrossRef] [PubMed]
28. Chanturia, V.A.; Dvoichenkova, G.P.; Koval'chuk, O.E.; Timofeev, A.S. Surface Composition and Role of Hydrophilic Diamonds in Foam Separation. *J. Min. Sci.* **2015**, *51*, 1235–1241. [CrossRef]
29. Zhong, B.; Zhang, J.Z.; Wang, H.Q.; Xia, L.; Wang, C.Y.; Zhang, X.D.; Huang, X.X.; Wen, G.W. Fabrication of novel silicon carbide-based nanomaterials with unique hydrophobicity and microwave absorption property. *Int. J. Appl. Ceram. Technol.* **2020**, *17*, 2598–2611. [CrossRef]
30. Wang, J.; Zhou, J.; Long, H.Y. Tribological, anti-corrosive properties and biocompatibility of the micro and nano-crystalline diamond coated Ti-6Al-4V. *Surf. Coat. Technol.* **2014**, *258*, 1032–1038. [CrossRef]
31. Feng, F.; Wu, Y.L.; Xin, H.T.; Chen, X.Q.; Guo, Y.Z.; Qin, D.Y.; An, B.L.; Diao, X.O.; Luo, H.W. Surface Characteristics and Biocompatibility of Ultrafine-Grain Ti after Sandblasting and Acid Etching for Dental Implants. *ACS Biomater. Sci. Eng.* **2019**, *5*, 5107–5115. [CrossRef] [PubMed]
32. McGuff, H.S.; Heim-Hall, J.; Holsinger, F.C.; Jones, A.A.; O'Dell, D.S.; Hafemeister, A.C. Maxillary osteosarcoma associated with a dental implant: Report of a case and review of the literature regarding implant-related sarcomas. *J. Am. Dent. Assoc.* **2008**, *139*, 1052–1059. [CrossRef] [PubMed]
33. Zhang, D.; Ren, L.; Zhang, Y.; Xue, N.; Yang, K.; Zhong, M. Antibacterial activity against Porphyromonas gingivalis and biological characteristics of antibacterial stainless steel. *Colloids Surf. B Biointerfaces* **2013**, *105*, 51–57. [CrossRef] [PubMed]
34. Nel, A.E.; Madler, L.; Velegol, D.; Xia, T.; Hoek, E.M.; Somasundaran, P.; Klaessig, F.; Castranova, V.; Thompson, M. Understanding biophysico-chemical interactions at the nano–bio interface. *Nat. Mater.* **2009**, *8*, 543–557. [CrossRef] [PubMed]

35. Monopoli, M.P.; Walczyk, D.; Campbell, A.; Elia, G.; Lynch, I.; Bombelli, F.B.; Dawson, K.A. Physical–chemical aspects of protein corona: Relevance to in vitro and in vivo biological impacts of nanoparticles. *Am. Chem. Soc.* **2011**, *133*, 2525–2534. [CrossRef] [PubMed]
36. Thomson, L.A.; Law, F.C.; Rushton, N.; Franks, J. Biocompatibility of diamond-like carbon coating. *Biomaterials* **1991**, *12*, 37–40. [CrossRef]

Article

Design of Ti-Zr-Ta Alloys with Low Elastic Modulus Reinforced by Spinodal Decomposition

Rong Wu [1], Qionghua Yi [1], Shan Lei [1,*], Yilong Dai [2] and Jianguo Lin [2,*]

[1] School of Materials Science and Engineering, Xiangtan University, Xiangtan 411105, China; 201921001406@smail.xtu.edu.cn (R.W.); yqh@aptmed.com (Q.Y.)
[2] Key Laboratory of Materials Design and Preparation Technology of Hunan Province, Xiangtan University, Xiangtan 411105, China; daiyilong@xtu.edu.cn
* Correspondence: leishan@xtu.edu.cn (S.L.); lin_j_g@xtu.edu.cn (J.L.)

Abstract: On the basis of the ternary phase diagram of Ti-Zr-Ta alloys and the d-electron orbital theory, the alloys with the nominal compositions of Ti-40Zr-20Ta (TZT1), Ti-35Zr-20Ta (TZT2) and Ti-30Zr-20Ta (TZT3) (in atom composition) were designed. The alloys were solution-treated (STed) at 1173 K for 3 h, and then aged (Aed) at 973 K for 6 h. The microstructure and mechanical properties of the three alloys were characterized. The results show that the three alloys comprised β-equiaxed grains after solution treatment at 1173 K for 3 h, and the β phases separated into $β_1/β_2$ phases by the spinodal decomposition in the alloys after being aged at 973 K for 6 h. The spinodal decomposition significantly promoted the yield strength of the alloys. For the TZT1 alloy, the yield strength increased from 1191 MPa (in the STed state) to 1580 MPa (in the Aed state), increasing by about 34%. The elastic moduli of the TZT1, TZT2 and TZT3 alloys after solution treatment at 1173 K were 75.0 GPa, 78.2 GPa and 85.8 GPa, respectively. After being aged at 973 K for 6 h, the elastic moduli of the three alloys increased to 81 GPa, 90 GPa and 92 GPa, respectively. Therefore, the spinodal decomposition can significantly promote the strength of the Ti-Zr-Ta alloys without a large increase in their elastic modulus.

Keywords: β-titanium alloy; spinodal decomposition; alloy design; elastic modulus; yield strength; microstructure

Citation: Wu, R.; Yi, Q.; Lei, S.; Dai, Y.; Lin, J. Design of Ti-Zr-Ta Alloys with Low Elastic Modulus Reinforced by Spinodal Decomposition. *Coatings* 2022, 12, 756. https://doi.org/10.3390/coatings12060756

Academic Editor: Alina Vladescu

Received: 7 May 2022
Accepted: 30 May 2022
Published: 31 May 2022

Publisher's Note: MDPI stays neutral with regard to jurisdictional claims in published maps and institutional affiliations.

Copyright: © 2022 by the authors. Licensee MDPI, Basel, Switzerland. This article is an open access article distributed under the terms and conditions of the Creative Commons Attribution (CC BY) license (https://creativecommons.org/licenses/by/4.0/).

1. Introduction

Titanium and its alloys are considered some of the most promising biomedical metal materials due their low density, high specific strength, low elastic modulus, good corrosion resistance, etc. [1–3]. Traditional titanium alloys, such as Ti-6Al-4V, normally contain Al and V elements, which have been proven to be biotoxic after implantation in the human body, and thus there is a risk of harm to human health [4,5]. On the other hand, as a human bone implant material, it should have a high strength and a low elastic modulus. Compared with other metal materials, titanium and its alloys have a relatively low elastic modulus, but the value is still much higher than that of human bones, which may cause the "stress shielding effect" when they are implanted into the human body, eventually leading to the failure of implants [6], Therefore, in recent years, a series of metastable β-titanium alloys have been developed for biomedical use by adding non-toxic beta-phase-stabilizing elements (such as Nb, Zr, Ta, Mo, Sn, etc.) to titanium [7,8]. These alloys not only have good biocompatibility, but also have an extremely low elastic modulus compared to conventional titanium alloys. For example, it has been reported that the elastic modulus of the Ti-Nb-Ta-Zr (TNTZ) alloy is about 50–55 GPa [9], and the Ti–24Nb–4Zr–8Sn alloy developed by Yang et al. has a lower elastic modulus, which reaches about 40 GPa [10].

It is well-documented that the decrease in the elastic modulus of titanium alloys can be realized through the design of alloy composition and the regulation of microstructure.

However, the decrease in the elastic modulus of an alloy usually leads to a reduction in its strength. Addressing the dilemma is a challenge in the material field.

A spinodal decomposition refers to a uniform phase transition, in which a solid solution decomposes into different phases with the same structure due to its instability [11,12]. The phase transition does not require nucleation, but rapidly forms two phases with the same crystal structure, which can increase the strength greatly without a significant increase in the elastic modulus of the alloys. Recently, Liu et al. calculated the isothermal cross section at 1173 K and 973 K of the phase diagram of a Ti-Zr-Ta ternary alloy system using the CALPHAD method, finding that the alloy system has a wide-ranging solid-solution gap at the two temperatures [13]. It allowed us to design β-titanium alloys with spinodal decomposition.

In light of it, in the present work, the Ti-Zr-Ta alloys with a low elastic modulus reinforced by spinodal decomposition were designed based on the isothermal cross section at 973 K of the phase diagram of a Ti-Zr-Ta ternary alloy system and d-orbital theory. The designed alloys were solution-treated (STed) at 1173 K, and then aged (Aed) at 973 K. The microstructure and mechanical properties of the alloys after solution and aging treatment were characterized.

2. Experimental Methods

2.1. Design of Ti-Zr-Ta Alloys with Spinodal Decomposition

Figure 1 is the isothermal cross section at 973 K of the phase diagram of the ternary Ti-Zr-Ta alloy, showing that there is a large solid-solution gap in the Ti-Zr-Ta alloy at 973 K. It implies that spinodal decomposition may occur in the alloys with compositions falling into the area of the solid-solution gap from the point view of dynamics. So, the alloys in present work were designed to have compositions within the region of the solid-solution gap.

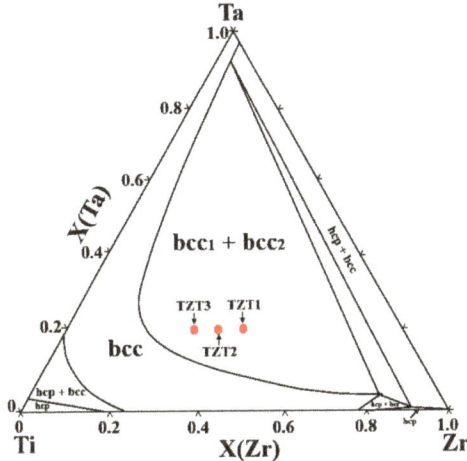

Figure 1. Isothermal cross section at 973K of the phase diagram of the ternary Ti-Zr-Ta alloy [13].

With regard to the modulus of the alloys, it has been reported that the d-orbital theory can be used to predict the elastic modulus of Ti-based alloys [14], in which two parameters, B_o and M_d, are defined, where B_o (the bond order) is a measure of the covalent bond strength between Ti and the alloying element, and M_d (the metal d-orbital energy level) correlates with the electronegativity and the metallic radius of elements. The average values of $\overline{B_o}$ and $\overline{M_d}$ are defined by taking the compositional averages of the parameters of B_o and M_d, respectively. Generally, an alloy with a low $\overline{B_o}$ and/or a high $\overline{M_d}$ may exhibit a low elastic modulus. So, the compositions of the designed alloys should give a low $\overline{B_o}$ and/or a high $\overline{M_d}$. Considering that Ta has a high melting point and high cost, we minimized Ta content when choosing the alloy composition. So, for the present work, we designed Ti-Zr-Ta alloys with compositions of Ti-40Zr-20Ta, Ti-35Zr-20Ta and Ti-30Zr-20Ta (in atom percent), which are represented by TZT1, TZT2 and TZT3, respectively. The values of $\overline{B_o}$ and $\overline{M_d}$ for the TNT1, TNT2 and TNT3 alloys were calculated, and the results are listed in Table 1.

Table 1. The calculated values of $\overline{B_o}$ and $\overline{M_d}$ for TZT1, TZT2 and TZT3 alloys.

Alloys	B_o	M_d
TZT1	2.9792	2.6586
TZT2	2.9644	2.6342
TZT3	2.9496	2.6099

2.2. Preparation of Ti-Zr-Ta Alloys

Titanium alloy ingots with the nominal compositions of Ti-40Zr-20Ta, Ti-35Zr-20Ta and Ti-30Zr-20Ta (in atom percent) were prepared by using a vacuum arc melting furnace. The purity of raw materials was above 99.9%. To obtain alloys with a uniform composition, the ingots were flipped and remelted more than 5 times. Then, the ingots used in subsequent experiments were obtained by suction-casting with a geometric size of 70 mm × 12 mm × 2 mm. The as-cast alloys were sealed in a vacuum quartz tube and placed into a muffle furnace for solution treatment (ST) at 1173 K for 3 h, and then quenched in an ice-cold saline solution. The ingots, after solution treatment, were sealed in a vacuum quartz tube again and placed in a muffle furnace for aging treatment at 973 K for 6 h, and then cooled with furnace cooling to room temperature.

2.3. Microstructure and Mechanical Properties' Characterization

The microstructure of the STed alloys was characterized using the optical microscopy, and the phase constitutions of the alloys after STed and STed + Aed were identified using X-ray diffraction (XRD) operated at 50 kV and 100 mA with Cu Kα radiation (λ = 1.5406 nm). The samples for the optical microscopic observations and XRD were cut from the ingots into small pieces of 2 mm × 2 mm × 1 mm. The samples were mounted and ground using a series of SiC sandpapers from 400 to 1200, and then polished with diamond suspension from 5 to 0.5 µm using ethyl alcohol as a lubricant. Transmission electron microscopic (TEM) observation was conducted on a JEM-2100 (JEOL Ltd., Tokyo, Japan) with an operating voltage of 160 KV, and TEM samples were prepared using Ar + ion milling.

Elastic moduli of the STed alloys and STed+Aed alloys were determined by Triobindenter TI 900 (Hysitron, Minneapolis, MN, USA) using monotonic load tests at a depth of up to 7000 nm using a Berkovich tip with a measured radius of 5000 nm. Tensile tests were carried out on an Instron 5569 (Instron, Norwood, MA, USA) universal testing machine, using the tensile samples with a gauge section of 1 mm × 2.5 mm × 8 mm, and the geometry of the samples is schematically shown in Figure 2.

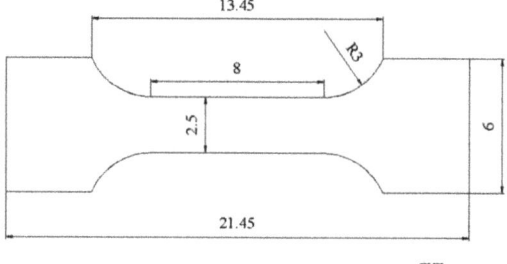

Figure 2. Geometric sketch of the tensile samples.

3. Results and Discussion

3.1. Microstructure

Figure 3 shows the optical images of the STed samples of TZT1, TZT2 and TZT3 alloys. The microstructure of the three alloys after solution treatment comprised equiaxed grains with the average grain size of 100 um. The results of XRD analysis reveal that only β phases existed in the STed samples, as shown in Figure 4, while after being aged at 973 K for 6 h, each diffraction peak of β phases on the XRD pattern split into two peaks, corresponding to β_1 and β_2 (see Figure 4), implying that the β phase separated into β_1 and β_2 during the aging treatment. To further confirm it, TEM observations were conducted on the STed+Aed samples. Figure 5 shows the TEM bright-field images of the STed+Aed samples, showing that the self-organized, tweed-like microstructure with a modulated contrast along [100] and [010] directions was formed in the alloys. The result further indicates that the spinodal decomposition occurred in the STed samples during the aging treatment at 973 K for 6 h. The alternating dark and bright regions in the modulated contrast microstructure, which could be described as sinusoidal composition modulations with a fixed wavelength [11,12], were within the order of ~10 nm in width; thus, the modulation wavelength in the alloys was ~10 nm.

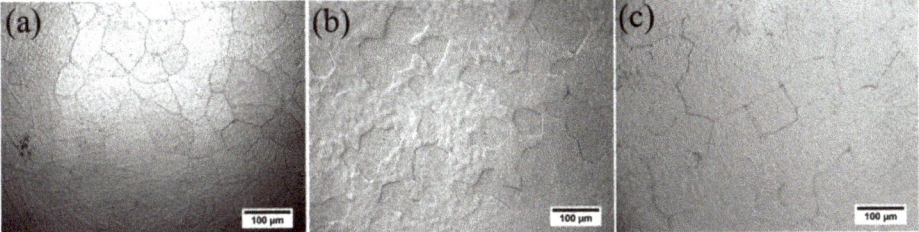

Figure 3. The optical images of the STed samples of (**a**) TZT1, (**b**) TZT2 and (**c**) TZT3 alloys.

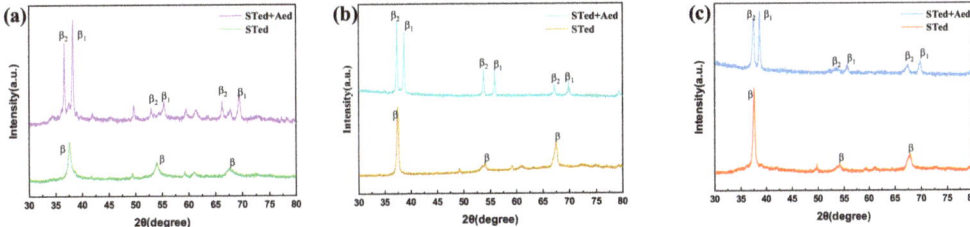

Figure 4. XRD patterns of the three alloys after solution treatment at 1173 K for 3 h and age treatment at 973 K for 6 h: (**a**) TZT1, (**b**) TZT2 and (**c**) TZT3.

3.2. Mechanical Properties

The mechanical properties of the alloys after being solution-treated and aged were evaluated by tensile tests at room temperature. Figure 6 illustrates the yield strengths ($\sigma_{0.2}$) of the STed and STed+Aed samples obtained from their stress–strain curves. It can be seen that the yield strengths of TZT1, TZT2 and TZT3 alloys after solution treatment were about 1191 MPa, 1150 MPa and 1075 MPa, respectively, while after being aged at 973 K for 6 h, the yield strengths of the three alloys were 1580 MPa, 1320 MPa and 1180 MPa, respectively. So, the aging treatment led to the significant increase in the yield strength of the three alloys. For TZT1, TZT2 and TZT3 alloys aged at 973 K for 6 h, the yield strength increased by about 34%, 15% and 10%, respectively, in comparison with the solution-treated alloys. Therefore, the modulated microstructure induced by spinodal decomposition had a strong strengthening effect, which may be responsible for the great increase in the yield strength of the three alloys.

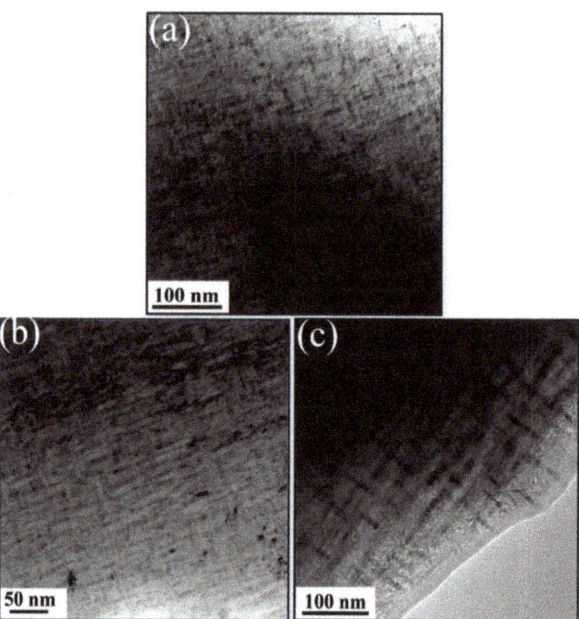

Figure 5. TEM bright-field images of the three alloys after being aged at 973 K for 6 h: (a) TZT1, (b) TZT2 and (c) TZT3.

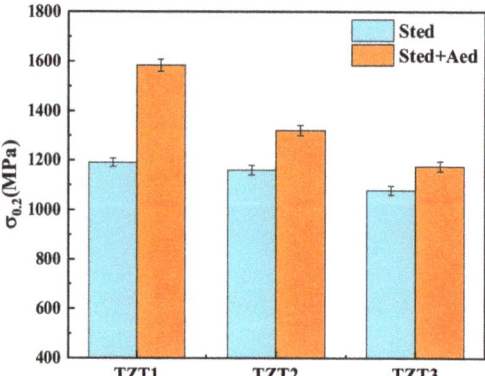

Figure 6. The yield strength of the TZT1 and TZT2 and TZT3 alloys after solution treatment at 1173 K for 3 h and age treatment at 973 K for 6 h.

Moreover, the elastic moduli of the alloys after solid-solution treatment and being aged were measured by the nanoindentation method, and the results are illustrated in Figure 7. For comparison, the elastic modulus of pure Ti is also illustrated in Figure 7. It can be seen that the elastic moduli of the three alloys after solid-solution treatment were 75.0 GPa, 78.2 GPa and 85.8 GPa, respectively, which are much lower than that of pure titanium (about 113 GPa). Generally, alloys with a low value of $\overline{B_0}$ and/or a high value of $\overline{M_d}$ may exhibit a low elastic modulus, according to the d-orbital theory [15–18]. From Table 1, one can see that the values of $\overline{B_0}$ for the three alloys were identical, but their values of $\overline{M_d}$ were in the order of TZT1 > TZT2 > TZT3. Therefore, the STed TZT1 alloy exhibited the lowest elastic modulus value due to having the highest $\overline{M_d}$.

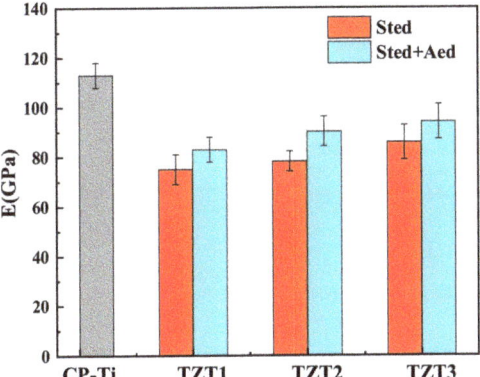

Figure 7. The elastic moduli of the CP-Ti and the TZT1 and TZT2 and TZT3 alloys after solution treatment at 1173 K for 3 h and being aged at 973 K for 6 h.

After being aged at 973 K for 6 h, the elastic moduli of the alloys were 81 GPa, 90 GPa and 92 GPa, as shown in Figure 7, increasing by about 8%, 15% and 7% in comparison with the corresponding STed alloys, respectively. As a result, spinodal decomposition largely increased the strength of the alloys without a significant increase in their elastic modulus.

4. Conclusions

The Ti-Zr-Ta alloys with the compositions of Ti-40Zr-20Ta, Ti-35Zr-20Ta and Ti-30Zr-20Ta (in atom percent) were designed based on the Ti-Zr-Ta phase diagram and d-orbital theory. The microstructure and mechanical properties of the three alloys were characterized. The main conclusions are as follows:

(1) The microstructure of the designed alloys consisted of β-equiaxed grains after solution treatment at 1173 K for 3 h. After being aged at 973 K for 6 h, spinodal decomposition occurred, forming the self-organized, tweed-like microstructure with a modulated contrast along [100] and [010] directions in the three alloys.
(2) The yield strength of the STed alloys of TZT1, TZT2 and TZT3 were 1191 MPa, 1150 MPa and 1075 MPa, respectively, after being aged at 973 K for 6 h. The yield strengths of the three alloys were increased to 1580 MPa, 1320 MPa and 1180 MPa, respectively. The spinodal strengthening may be responsible for the increase in the strength of the alloys.
(3) The elastic moduli of the three alloys after solid-solution treatment were 75.0 GPa, 78.2 GPa and 85.8 GPa, respectively, which was much lower than that of pure titanium (about 113 GPa). The TZT1 alloy exhibited the lowest elastic modulus due to its high value of $\overline{M_d}$.
(4) After being aged at 973 K for 6 h, the elastic moduli of the three alloys slightly increased. Spinodal decomposition can significantly promote the strength of the Ti-Zr-Ta alloys without a large increase in their elastic modulus.

Author Contributions: Conceptualization, R.W. and S.L.; methodology, R.W. and Q.Y.; Data curation, Q.Y. and Y.D.; Writing—original draft preparation, R.W.; Writing—review and editing, S.L. and J.L.; Supervision, S.L. and J.L. All authors have read and agreed to the published version of the manuscript.

Funding: This research was financially supported by the National Natural Science Foundation of China (Grant No. 11872053 and No. 51971190) and the Project of Hunan Provincial Education Department (Grant No. 21C0069).

Institutional Review Board Statement: Not applicable.

Informed Consent Statement: Not applicable.

Data Availability Statement: Not applicable.

Conflicts of Interest: The authors declare no conflict of interest.

References

1. Guo, L.; Naghavi, S.A.; Wang, Z.; Varma, S.N.; Han, Z.; Yao, Z.; Wang, L.; Wang, L.; Liu, C. On the design evolution of hip implants: A review. *Mater. Des.* **2022**, *216*, 110552. [CrossRef]
2. Besinis, A.; Hadi, S.D.; Le, H.; Tredwin, C.; Handy, R. Antibacterial activity and biofilm inhibition by surface modified titanium alloy medical implants following application of silver, titanium dioxide and hydroxyapatite nanocoatings. *Nanotoxicology* **2017**, *11*, 327–338. [CrossRef] [PubMed]
3. Zhang, L.C.; Chen, L.Y. A review on biomedical titanium alloys: Recent progress and prospect. *Adv. Eng. Mater.* **2019**, *21*, 1801215. [CrossRef]
4. Kyzioł, K.; Kaczmarek, Ł.; Brzezinka, G.; Kyzioł, A. Structure, characterization and cytotoxicity study on plasma surface modified Ti–6Al–4V and γ-TiAl alloys. *Chem. Eng. J.* **2014**, *240*, 516–526. [CrossRef]
5. Wang, X.; Li, Y.; Xiong, J.; Hodgson, P.D. Porous TiNbZr alloy scaffolds for biomedical applications. *Acta Biomater.* **2009**, *5*, 3616–3624. [CrossRef] [PubMed]
6. Geetha, M.; Singh, A.K.; Asokamani, R.; Gogia, A.K. Ti based biomaterials, the ultimate choice for orthopaedic implants—A review. *Prog. Mater. Sci.* **2009**, *54*, 397–425. [CrossRef]
7. Welsch, G.; Boyer, R.; Collings, E. *Materials Properties Handbook: Titanium Alloys*; ASM International: Almere, The Netherlands, 1993.
8. Abigail, M.N. The Design and Development of a Novel Beta Ti Alloys with Low Elastic Modulus for Biomedical Application. Master's Thesis, University of Johannesburg, Johannesburg, South Africa, 2020.
9. Hagihara, K.; Nakano, T.; Maki, H.; Umakoshi, Y.; Niinomi, M. Isotropic plasticity of β-type Ti-29Nb-13Ta-4.6 Zr alloy single crystals for the development of single crystalline β-Ti implants. *Sci. Rep.* **2016**, *6*, 29779. [CrossRef] [PubMed]
10. Yang, Y.; Castany, P.; Cornen, M.; Prima, F.; Li, S.; Hao, Y.; Gloriant, T. Characterization of the martensitic transformation in the superelastic Ti–24Nb–4Zr–8Sn alloy by in situ synchrotron X-ray diffraction and dynamic mechanical analysis. *Acta Mater.* **2015**, *88*, 25–33. [CrossRef]
11. Langer, J.S. Theory of spinodal decomposition in alloys. *Ann. Phys.* **1971**, *65*, 53–86. [CrossRef]
12. Tang, Y.; Goto, W.; Hirosawa, S.; Horita, Z.; Lee, S.; Matsuda, K.; Terada, D. Concurrent strengthening of ultrafine-grained age-hardenable Al-Mg alloy by means of high-pressure torsion and spinodal decomposition. *Acta Mater.* **2017**, *131*, 57–64. [CrossRef]
13. Wang, J.L. Study on phase diagram, thermodynamics of phase equilibrium and alloy design of titanium alloy. Doctoral Dissertation, Central South University, Changsha, China, 2015. (Unpublished).
14. Zhang, D.; Lin, J.; Jiang, W.; Ma, M.; Peng, Z. Shape memory and superelastic behavior of Ti–7.5 Nb–4Mo–1Sn alloy. *Mater. Des.* **2011**, *32*, 4614–4617. [CrossRef]
15. Abdel-Hady, M.; Hinoshita, K.; Morinaga, M. General approach to phase stability and elastic properties of β-type Ti-alloys using electronic parameters. *Scr. Mater.* **2006**, *55*, 477–480. [CrossRef]
16. Sung, H.H.; Sang, W.P.; Chan, H.P.; Jong, T.Y.; Ki, B.K. Relationship between phase stability and mechanical properties on near/metastable β-type Ti–Cr-(Mn) cast alloys. *J. Alloys Compd.* **2020**, *821*, 153516.
17. Bolzoni, L.; Alqattan, M.; Yang, F.; Peters, L. Design of β-eutectoid bearing ti alloys with antibacterial functionality. *Mater. Lett.* **2020**, *278*, 128445. [CrossRef]
18. Shao, L.F.; Du, Y.H.; Dai, K.; Wu, H.; Wang, Q.G.; Liu, J.; Tang, Y.J.; Wang, L.Q. β-Ti alloys for orthopedic and dental applications: A review of progress on improvement of properties through surface modification. *Coatings* **2021**, *11*, 1446. [CrossRef]

Article

Preparation of Ti-Al-Si Gradient Coating Based on Silicon Concentration Gradient and Added-Ce

Zihan Wang, Faguo Li *, Xiaoyuan Hu, Wei He, Zhan Liu and Yao Tan

School of Materials Science and Engineering, Xiangtan University, Xiangtan 411105, China; wangzihann0524@163.com (Z.W.); hxyajr@outlook.com (X.H.); 15750022348@163.com (W.H.); liuzhanxtu@yeah.net (Z.L.); aptx486912010504@163.com (Y.T.)
* Correspondence: lifaguo@xtu.edu.cn

Abstract: Titanium and titanium alloys have excellent physical properties and process properties and are widely used in the aviation industry, but their high-temperature oxidation resistance is poor, and there is a thermal barrier temperature of 600 °C, which limits their application as high-temperature components. The Self-generated Gradient Hot-dipping Infiltration (SGHDI) method is used to prepare the Ti-Al-Si gradient coating based on the silicon concentration with a compact $Ti(Al,Si)_3$ phase layer, which can effectively improve the high-temperature oxidation resistance of the titanium alloy. Adding cerium can effectively inhibit the generation of the τ_2: $Ti(Al_xSi_{1-x})_2$ phase within a certain hot infiltration time so as to form a continuous dense Al_2O_3 layer to further improve the oxidation resistance of the coating. Studies have found that multiple Ti-Al binary alloy phase layers are formed during the high-temperature oxidation process, which has the effect of isolating oxygen and crack growth, and effectively improving the high-temperature resistance of the coating oxidation performance.

Keywords: titanium alloy; Ti-Al-Si gradient coating; cerium; self-generated gradient hot-dipping infiltration (SGHDI) method; microstructure and properties

Citation: Wang, Z.; Li, F.; Hu, X.; He, W.; Liu, Z.; Tan, Y. Preparation of Ti-Al-Si Gradient Coating Based on Silicon Concentration Gradient and Added-Ce. *Coatings* 2022, 12, 683. https://doi.org/10.3390/coatings12050683

Academic Editor: Ana-Maria Lepadatu

Received: 9 April 2022
Accepted: 14 May 2022
Published: 16 May 2022

Publisher's Note: MDPI stays neutral with regard to jurisdictional claims in published maps and institutional affiliations.

Copyright: © 2022 by the authors. Licensee MDPI, Basel, Switzerland. This article is an open access article distributed under the terms and conditions of the Creative Commons Attribution (CC BY) license (https://creativecommons.org/licenses/by/4.0/).

1. Introduction

High-temperature titanium alloys are important structural materials for the key components of modern aero-engines, which are known as the 'pearl in the crown' of industrial manufacturing. Aero-engine is a concentrated embodiment of a country's comprehensive national strength, industrial foundation and technological level, and it is an important guarantee of national security. As the aviation industry develops rapidly, the thrust-to-weight ratio of an areo-engine is required to be higher. Titanium alloys can be used in fan and compressor parts of aero-engines instead of Ni-based superalloys, such as compressor discs, vanes, navigations and adapter rings, which can reduce the weight of the compressor by 30–35% [1]. At present, the amount of titanium alloys has accounted for 25–40% of the total weight of advanced aero-engines. However, traditional titanium alloys have to be used in air for a long time at the temperature of 540–600 °C or so, above which will result in excessive oxidation and oxygen brittleness caused by oxygen diffusion to the substrate [2]. At present, the maximum short-term operating temperature of mature high-temperature titanium alloys is 600–750 °C. As the temperature rises, the high-temperature creep and high-temperature oxidation resistance of titanium alloys will significantly decrease, which has become a bottleneck restricting the development of titanium alloys' application at higher temperatures [3]. Thermal corrosion is a phenomenon in which oxygen and other corrosive gases work together with salts deposited on the surface of materials in a high-temperature environment to accelerate corrosion, and its harmfulness is much greater than thermal oxidation. In addition, the research on high-temperature thermal corrosion resistance of high-temperature coatings in service above 1000 °C will be the focus of research in the future [4]. Therefore, in order to further improve the thrust–weight ratio of aero-engines and meet the requirements of materials for high-performance aero-engines, it is imperative to develop a new high-temperature titanium alloy coating with higher operating temperature.

For decades, people have studied coating elements such as Al, Si, Cr, Nb, C, S and Mo, and found that aluminum was the most effective among many anti-oxidation elements [5]. Aluminum reacts at high temperatures to produce stable Al_2O_3 [6], which has an excellent protective effect on the metal substrate. Therefore, aluminum is the preferred element for high-temperature oxidation resistance coatings [7]. Among the Ti–Al system's intermetallic compounds, only the $TiAl_3$ can form dense Al_2O_3 film in the air, whose Al content is 75% and oxidation resistance is good [8]. High-temperature diffusion aluminizing of titanium alloys can form a $TiAl_3$-rich coating, which can greatly improve its oxidation resistance [9]. The commonly used diffusion method is the hot-dipping method, which was initially applied to steel materials [10] and is often used to improve the corrosion resistance, abrasive resistance and oxidation resistance at high temperature. Studies have found that the hot-dipping method can also improve the performance of titanium alloys [11–13]. However, aluminum hot-dipping cannot provide effective protection for titanium alloy above 800 °C. In addition, because multiple Ti-Al phase layers are generated, there will be penetrating cracks because the thermal expansion coefficient of the phase layer does not match the substrate [11]. However, adding silicon can reduce the number of transverse cracks in the coating. Compared with the hot-dip aluminizing layer, the Ti-al-si layer is tough and compact [14]. The oxidation resistance temperature of the coating can be increased to 800–850 °C [15,16]. However, it is difficult to form the $TiAl_3$ phase layer by directly adding silicon because Ti and Si tend to react first to generate the Ti-Si intermediate phase [17], thus changing the reaction path between Ti, Al and Si. Therefore, figuring out how to introduce silicon after forming the $TiAl_3$ phase layer preferentially has become a key problem in optimizing the Al-Si coating of titanium alloys.

Therefore, this paper proposes a Self-generated Gradient Hot-dipping Infiltration (SGHDI) method to achieve the Ti-Al-Si gradient coatings after the formation of the $TiAl_3$ phase layer preferentially and the introduction of silicon [18]. The characteristics of the coatings are as follows: the Ti-Al-Si multiphase layer structure based on a Si concentration gradient is formed from the substrate to the outside: The dense $Ti(Al,Si)_3$ alloy layer of solid solution Si atoms, the dispersed bulk of τ_2: $Ti(Al_xSi_{1-x})_2$ phase + L-(Al,Si) phase and L-(Al,Si) phase. However, a dispersed τ_2 phase results in the non-denseness of the Al_2O_3 layer during a high-temperature oxidation process, which reduces the high-temperature oxidation resistance of the coating.

Adding rare earth elements (Ce, La, etc.) is considered to be an effective means to improve the high-temperature oxidation resistance and corrosion resistance of the coating [19]. Ce and La are enriched at grain boundaries, which can effectively reduce the high-temperature oxidation rate [20]. Rare earth elements can also refine the microstructure and structure of coatings [21]. Therefore, adding cerium is proposed to suppress the generation of the τ_2 phase and improve the high-temperature oxidation resistance of the coating. The effect of hot infiltration time on the microstructure and morphology of the Ti-Al-Si gradient coating added with Ce and the high-temperature oxidation resistance of the sample with excellent coating structure were studied.

2. Coating Preparation Method and Experiment Method

2.1. Coating Preparation Method

The essence of hot-dipping is a liquid/solid diffusion coupling reaction. When the titanium alloy is hot-dip aluminized, a Ti-Al binary intermetallic compound phase layer will be formed on the surface of the titanium alloy. When Si is added, it is difficult to form the $TiAl_3$ phase with excellent high-temperature oxidation resistance because the Ti atoms combine with Si atoms preferentially to form a Ti-Si binary intermediate phase. According to the ternary phase diagram of Ti-Al-Si, the Si atoms can solubilize in the $TiAl_3$ phase and form the secondary solid solution of $Ti(Al,Si)_3$ [22]. Therefore, the $Ti(Al,Si)_3$ phase can be used as the medium for Si atoms to dissolve into the phase layer of Ti-Al alloy. Therefore, it is necessary to develop a new coating preparation method: the TiAl$_3$ phase layer is formed on the surface of the titanium alloy by hot-dip aluminizing, and then Si atoms are added into aluminum melt and Si atoms are dissolved into the TiAl$_3$ phase layer. Obviously, Si atoms will form a concentration gradient distribution in the TiAl$_3$ phase layer. Then, the key issue is how to introduce Si sources. We use the phenomenon that the aluminum melt can react with quartz glass to generate Si atoms and use the quartz glass tube as a container for pure aluminum liquid, thus cleverly solving the Si sources problem. The new coating preparation method is called the Self-generated Gradient Hot-dipping Infiltration (SGHDI) method.

2.2. Experiment Method

A 99.995 wt.% pure Al, Al-20 wt.% Ce master alloy, high-purity quartz glass tube (providing active Si atoms source and acting as a container), and 10-mm-diameter Ti-6Al-4V (TC4) alloy rod were selected for the experiment. The hot infiltration process is as follows:

(1) Degrease the surface of the TC4 alloy with a metal cleaning agent and dry it;
(2) Melt the high-purity aluminum in the vertical pit furnace;
(3) Pour the high-purity aluminum melt in step (2) into the high-purity quartz tube, keep the melt state at a high-temperature of 800 °C, immerse the TC4 alloy into the high-purity aluminum melt, hold for a certain time, quickly extract from the melt, and then quench.

The Al-1 wt.% Ce solution was prepared by adding a moderate amount of Al-20 wt.% Ce master alloy into the high-purity aluminum melt so as to achieve the purpose of adding Ce.

The high-temperature oxidation experiment involves placing coated/uncoated specimens, respectively, in small corundum crucibles, weighing, recording, and then putting them in the box furnace (Xiangtan Samsung Instrument Co., LTD, Xiangtan, China), heating up to 800 °C, taking out a specimen at regular intervals, weighing and recording in order to calculate how much weight is added per unit area at different times, and obtaining the high-temperature oxidation weight curves of coated/uncoated specimens.

The microstructure, phase constitute, and element distribution of the coating were characterized and measured by SEM (ZEISS EVO MA10, Zeiss, Jena, Germany), EDS (OXFORD X-MAXN, Zeiss, Jena, Germany).

3. Results and Discussion

3.1. Microstructure Characterization of Unadded Cerium Coating

Figure 1 shows the microstructures of the Al-Si coating of TC4 at 800 °C for different hot-dipping times. The dipping coating has obvious delamination, which can be divided into two regions: (1) the surface layer is a liquid phase, *L*-(Al, Si); (2) the alloy phase layer of mutual diffusion between substrate and liquid phase: the dense alloy phase layer, which metallurgically combined with the substrate, and the dispersed bulk alloy phase layer.

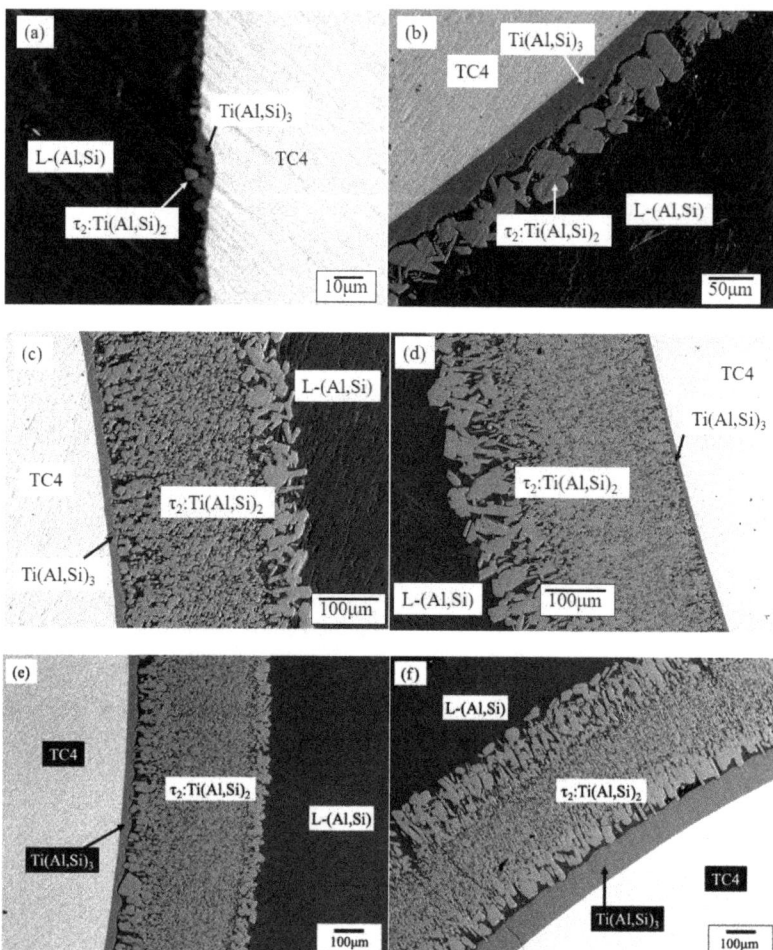

Figure 1. The microstructures of Ti-Al-Si gradient coatings on TC4 alloy at 800 °C, (**a**) 10 min, (**b**) 30 min, (**c**) 50 min, (**d**) 80 min, (**e**) 90 min, (**f**) 110 min.

The EDS results show that the elemental atomic ratio is within the range of Ti(Al,Si)$_3$ phase (Al:Si:Ti = 59:14:27) in the dense alloy phase layer, so it is a Ti(Al,Si)$_3$ phase. The EDS results show that the elemental atomic ratio is within the range of the τ_2: Ti(Al,Si)$_2$ phase (Al:Si:Ti = 55:12:33) in the dispersed bulk alloy phase layer, so it is the τ_2: Ti(Al,Si)$_2$ phase. In other words, the Ti(Al,Si)$_3$ phase, τ_2 + L-(Al,Si) phase and L-(Al,Si) phase are formed in sequence on the surface of titanium alloy.

Figure 1a shows that although the Ti(Al,Si)$_3$ phase layer is only a few microns in the early stage of the hot-dip aluminizing, it shows that this method realized the preferential reaction between Ti and Al to generate the TiAl$_3$ phase layer. Moreover, once the TiAl$_3$ phase layer is formed, a barrier is constructed to isolate Si atoms from directly contacting the surface of titanium alloy, and the reaction between Ti and Si is effectively prevented. In the subsequent long hot-dipping process (Figure 1a–f), although Si atoms can form an Ti(Al,Si)$_3$ phase by dissolving into an TiAl$_3$ phase, the diffusion rate of Al atoms in the liquid phase through the TiAl$_3$ phase layer to the surface of titanium alloy is much faster than that of Si atoms through the TiAl$_3$ phase layer. Therefore, the formation of a dense and thick Ti(Al,Si)$_3$ phase layer is guaranteed.

The morphology of the τ_2 phase layer is very interesting. The bulk τ_2 phase is uniform in the microstructure of hot-dipping coating (Figure 1a,b). However, with the increase in hot-dipping time, the τ_2 phase layer creates the stratification phenomenon of bulk size scale: inner large block τ_2, middle tiny block τ_2, and outer large block τ_2. The loose bulk τ_2 phase destroys the compactness of the Al_2O_3 aluminum layer, and the sharp corner also becomes the source of cracks, so it may adversely affect the overall oxidation-resistance of the coating.

3.2. The Microstructure Characterization of Added Cerium Coating

The microstructures of Ce-added hot infiltration coating are shown in Figure 2. With the increase in hot infiltration time, the thickness of the alloy phase layer gradually increased. When the hot infiltration time is less than 50 min, there is a very dense alloy layer, and the liquid phase covers it in the coating. The dense layer is the $Ti(Al,Si)_3$ phase layer confirmed by energy spectrum analysis. When the hot infiltration time reaches 60 min or even higher, the longer the hot infiltration time, the more the bulk phase appears in the liquid phase, which is the τ_2 phase by energy spectrum analysis. The $Ti(Al,Si)_3$ phase layer is gradually thinned due to the appearance of the τ_2 phase. It can be seen that the addition of Ce can suppress the generation of the τ_2 phase in a certain time range.

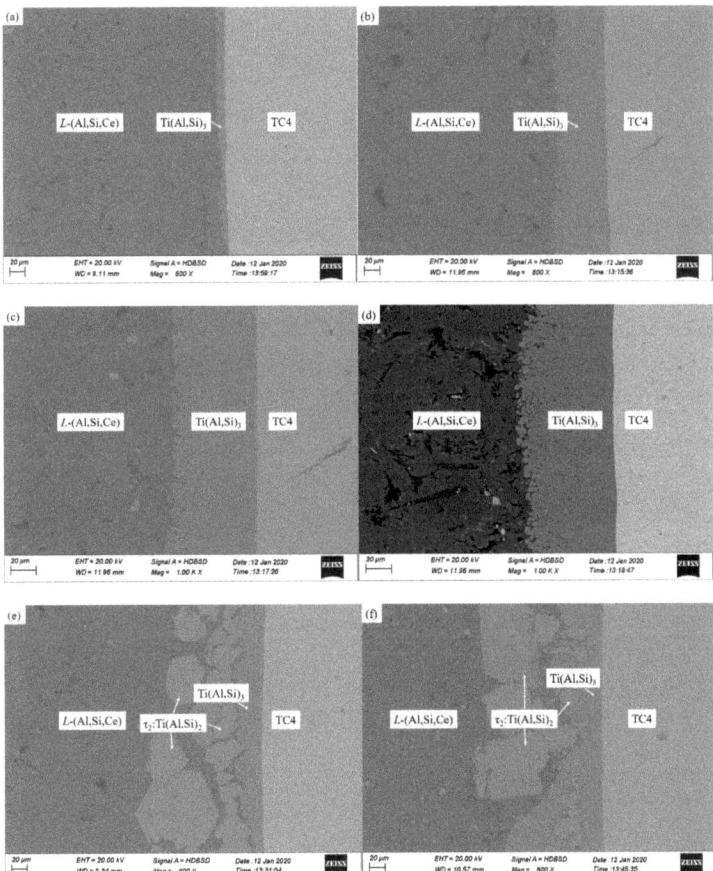

Figure 2. The microstructure of the Ce-added coating at different hot infiltration times, (**a**) 20 min; (**b**) 30 min; (**c**) 40 min; (**d**) 50 min; (**e**) 60 min; (**f**) 80 min.

The dense Ti(Al,Si)$_3$ alloy phase layer has excellent resistance to high-temperature oxidation and is the only single phase alloy phase layer in the hot infiltration process. Revealing the growth kinetics of the Ti(Al,Si)$_3$ alloy phase layer is helpful in guiding the structural design of the coating and adjust the high-temperature oxidation resistance of the coating. Considering that the thickness of the Ti(Al,Si)$_3$ alloy phase decreases instead of increasing after the hot infiltration time exceeds 50 min, this paper only analyzes the relationship between the thickness and time of the Ti(Al,Si)$_3$ alloy phase layer within 50 min. There exists an empirical formula [23]:

$$\delta = kt^n \tag{1}$$

where δ is the thickness (μm), t is time (min), k is the rate constant, and n is the kinetic exponent. The $n \leq 0.5$ means diffusion-controlled growth, and $n > 0.5$ means reaction-controlled growth.

The growth kinetics curve of the Ti(Al,Si)$_3$ alloy layer was fitted by Formula (1) (as shown in Figure 3):

$$\delta = 0.216t^{1.475} \tag{2}$$

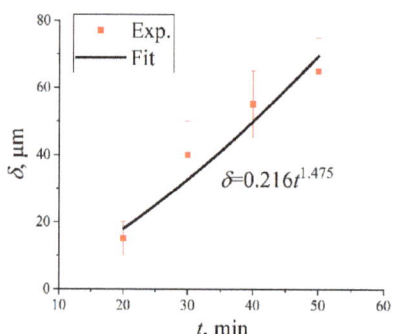

Figure 3. The growth kinetics curve of the Ti(Al,Si)$_3$ alloy layer.

The results show that it is reaction-controlled growth.

3.3. The Microstructure Characterization of Added Cerium Coating

No Ce compounds are found in the Ti-Al-Si gradient coating, but the effect of adding Ce is obvious, and the growth of the τ_2 phase is inhibited within 50 min of hot infiltration. According to the formation sequence of various phases, the coating growth stage can be divided into four important stages (Figure 4): the formation of the TiAl$_3$ phase, the growth of the TiAl$_3$ phase and Si atom solution, the growth of the Ti(Al,Si)$_3$ phase, and the formation and growth of the τ_2 phase.

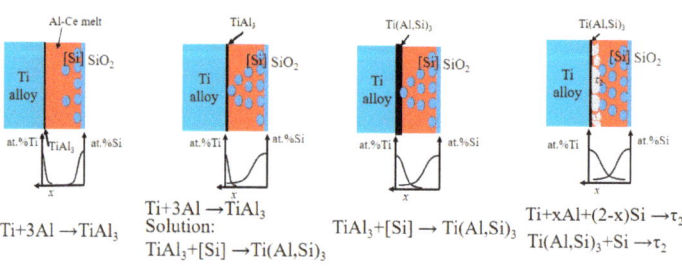

Figure 4. Schematic diagram of the growth mechanism of cerium-added coating.

When the TC4 alloy is immersed into Al-Ce melt, titanium atoms in TC4 diffuse to Al melt, and the TiAl$_3$ phase is formed firstly at the interface. The reaction equation is:

$$\text{Ti} + 3\text{Al} \rightarrow \text{TiAl}_3 \tag{3}$$

The Al melt on the quartz tube's wall reacts with SiO$_2$ at the same time, and the generated Si atoms diffuse toward TC4. As the diffusion time increases, solid Si atoms dissolve into TiAl$_3$, and the Ti(Al,Si)$_3$ phase is formed. The solution process is:

$$\text{TiAl}_3 + 3[\text{Si}] \rightarrow \text{Ti(Al,Si)}_3 \tag{4}$$

When Si in TiAl$_3$ is saturated, the Ti, Al and Si atoms at the front of the solid–liquid interface will react to form the Ti(Al,Si)$_2$ phase (τ_2 phase). In addition, the saturated Ti(Al,Si)$_3$ phase reacts with Si atoms to form the τ_2 phase [24]. The τ_2 phase is formed continuously, and the longer the diffusion time, the thicker the τ_2 phase layer and the thinner the Ti(Al,Si)$_3$ phase. The reaction equation to form τ_2 is:

$$\text{Ti} + x\text{Al} + (2 - x)\text{Si} \rightarrow \tau_2 \tag{5}$$

$$\text{Ti(Al,Si)}_3 + \text{Si} \rightarrow \tau_2 \tag{6}$$

3.4. High-Temperature Oxidation Resistance

The results can be obtained by fitting the weight curves of uncoated TC4 substrate alloy samples and the 60 min Ti-Al-Si hot-dipping samples by an isothermal oxidation comparison test at 800 °C in static air,

$$\Delta m = 2.107 t^{0.7} \text{ (uncoated)} \tag{7}$$

$$\Delta m = 0.624 t^{0.5} \text{ (Ti-Al-Si gradient coating)} \tag{8}$$

where Δm is the amount of oxidation weight gain (mg·cm$^{-1/2}$), and t is the oxidation time (h). It can be known from Formulas (7) and (8) that the mass increase rate of the uncoated samples is obviously more rapid than that of the coated samples. The high-temperature oxidation resistance of the Ti-Al-Si coating is better than that of the pure aluminum coating [13], and about the same as the oxidation resistance of the Al-Si coating prepared by Zhou W, et al. [25] by the low oxygen partial pressure self-fusing method. Therefore, this hot-dipping Ti-Al-Si gradient coating has excellent high-temperature oxidation resistance.

Figure 5a shows the isothermal oxidation microstructure of the 60 min hot-dipping sample. The outermost phase layer of the sample after the oxidation is τ_2: the Ti(Al,Si)$_2$ + α-Al$_2$O$_3$ phase. The loose τ_2 phase distribution causes the α-Al$_2$O$_3$ layer to become incompact. In the isothermal oxidation process, Al and Si atoms in the Ti(Al,Si)$_3$ phase layer diffuse to the substrate and react with Ti atoms to form some new dense phase layers, as shown in Figure 5b. Through energy spectrum analysis, Ti$_3$Al, TiAl, and Ti$_5$Si$_3$ are newly formed. From the distribution of these new phase layers, the diffusion rate of the Al atom is obviously higher than that of the Si atom. The Si atoms are mainly distributed in the original Ti(Al,Si)$_3$ layer, and part of the Si atoms react with Ti atoms to generate Ti$_5$Si$_3$ [26–28].

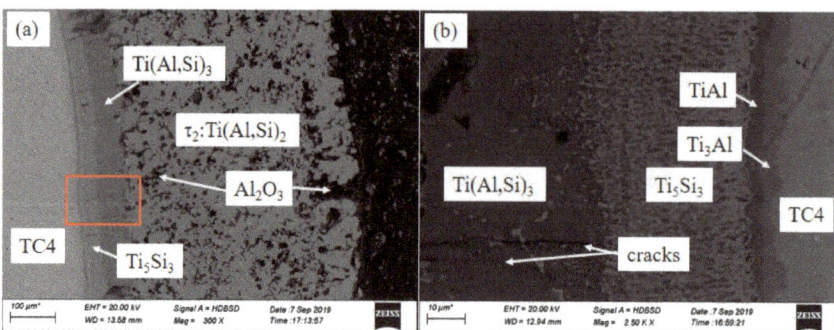

Figure 5. (**a**) Microstructure of the pure Ti-Al-Si gradient coating after oxidation for 120 h, (**b**) local magnification of red wireframe in (**a**).

After oxidation for 120 h, microcracks appeared in the Ti(Al,Si)$_3$ phase layer. However, the newly formed dense layer effectively prevented the crack propagation and prevented oxygen from entering the TC4 substrate, which has a good protective effect on the TC4 substrate.

Considering that the hot infiltration time to inhibit the generation of the τ_2 phase is less than 50 min, and the thickness of the Ti(Al,Si)$_3$ phase layer is less than 20 μm when the hot infiltration time is less than 20 min, and the long hot infiltration time is not suitable for industrial production, the Ti-Al-Si coating formed by a hot infiltration time of 30 min is more appropriate. In this paper, the sample with 30 min hot infiltration time was selected for the high-temperature oxidation experiment to analyze its high-temperature oxidation resistance. Figure 6a shows the experiment results of isothermal oxidation for 72 h cerium-added hot-dipping sample for 30 min in static air at 800 °C. The oxidized coating has seven distinct phase layers, and there are many white granular phases in the sixth and seventh phase layers. EDS (shown in Table 1) shows that Ti:Al = 3:1 in the 1# phase layer, so it is Ti$_3$Al. In the 2# phase layer, Ti:Al = 2:1, so it is Ti$_2$Al. In the 3# phase layer, Ti:Al = 1:1, so it is TiAl. In the 4# phase layer, Ti:Al = 1:2, so it is TiAl$_2$. The 5# phase layer is different from 1–4# phase layers in that there exists an oxygen element, and Ti:Al = 1:3, so it is the TiAl$_3$ phase layer. There are only Al, Si, Ti and O elements in the 6# phase layer, which are Ti(Al,Si)$_3$ and Al$_2$O$_3$ phases. There are five elements, Al, Si, O, Ti and Ce in the 7# phase layer, among which 85.34 at.% Al, 10.01 at.% O, 4.35 at.% Si, 0.09 at.% Ti and 0.21 at.% Ce in the gray area. The substrate of the 7# phase layer is L-(Al, Si, Ce) + Al$_2$O$_3$. The Al, Si, O, Ti and Ce in the white 8# phase layer are 57.17, 14.08, 3.77, 8.77 and 16.21 at.%, respectively, indicating that it is an Al$_6$Si$_2$Ce$_2$Ti compound. In addition, although cracks can run through the liquid phase layer and Ti(Al,Si)$_3$ layer, they will still be blocked by the newly formed Ti-Al binary layer and cannot penetrate the entire coating.

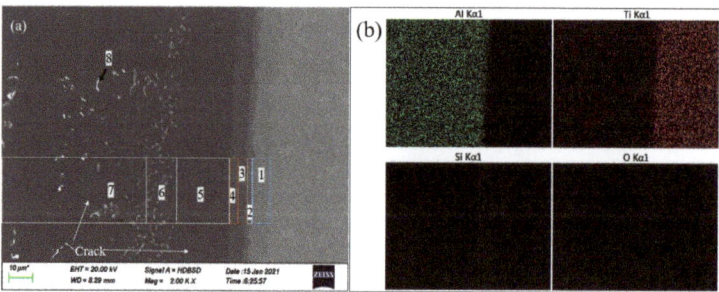

Figure 6. Microstructure (**a**) and EDS mapping date (**b**) of adding a cerium Ti-Al-Si gradient coating after 72 h oxidation.

Table 1. EDS data and main compounds of each alloy phase layer.

Phase Layer in Figure 6	Ti	Al	Si	O	Ce	Compounds
1#	76.46	23.54	0	0	0	Ti_3Al
2#	64.76	35.24	0	0	0	Ti_2Al
3#	47.40	52.60	0	0	0	$TiAl$
4#	34.47	65.53	0	0	0	$TiAl_2$
5#	25.43	67.07	0	7.50	0	$TiAl_3$
6#	24.48	54.90	11.95	8.67	0	$Ti(Al,Si)_3 + Al_2O_3$
7#	0.09	85.34	4.35	10.01	0.21	$L\text{-}(Al, Si, Ce) + Al_2O_3$
8#	8.77	57.17	14.08	3.77	16.21	$TiAl_6Si_2Ce_2$

Figure 6b shows the distribution of the elements in Figure 6a. The Al element is mainly distributed in the coating, while the Ti element has a small distribution in the coating, indicating that the Ti element has diffused into the coating. There are trace distribution of Si and O elements in the 5#–7# phase layer, indicating that the Ti-Al-Si coating added with Ce inhibits the formation of the loose τ_2 phase layer and greatly improves the high-temperature oxidation resistance.

Although Al, Si and Ti elements all diffuse each other during the high-temperature oxidation process, compared with the Ti-Al-Si coating without Ce added, first, there is no binary Ti-Si phase in the $Ti(Al,Si)_3$ phase layer, and the $Ti(Al,Si)_3$ phase layer remains dense. Second, $Al_6Si_2Ce_2Ti$ compounds are precipitated from the $L\text{-}(Al, Si, Ce)$ alloy layer during the high-temperature oxidation process, and the concentration of Si, Ce and Ti is much higher than the concentration of its own elements in the surrounding liquid phase, which restrains the generation of the τ_2 phase. Third, the continuous liquid layer facilitates further oxidation into a continuous Al_2O_3 phase layer.

3.5. The Formation Mechanism of High-Temperature Oxidation Microstructure of Adding Ce Coatings

Figure 7 is a schematic diagram of the high-temperature oxidation mechanism of Ce-added Ti-Al-Si gradient coating with 30 min hot infiltration. Before high-temperature oxidation, the coating consisted of a dense $Ti(Al,Si)_3$ phase layer and a liquid phase. After high-temperature oxidation, oxygen diffuses toward the substrate, forming dense Al_2O_3 in the liquid phase, the Al atoms in the $Ti(Al,Si)_3$ diffuse toward the substrate and form a new dense alloy phase layer $TiAl_3$. When the high-temperature oxidation time increases, several new Ti-Al alloy phase layers (Ti_3Al, Ti_2Al, $TiAl$, $TiAl_2$) are formed between $TiAl_3$ and the substrate, which can effectively prevent the diffusion of oxygen and crack towards the substrate.

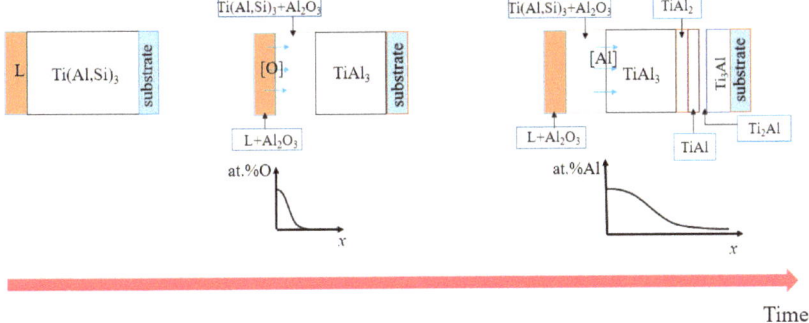

Figure 7. Schematic diagram of the high-temperature oxidation mechanism.

4. Conclusions

In this paper, a Ti-Al-Si gradient coating was successfully prepared by a self-generated gradient hot-dipping method, and the generation of τ_2 was suppressed by adding Ce. The formation mechanism and high-temperature oxidation resistance of the coating were discussed. The main conclusions are as follows:

(1) The thickness of the Ti(Al,Si)$_3$ phase with a Ce-added Ti-Al-Si gradient coating increases gradually with the increase in hot infiltration time, and the τ_2: Ti(Al,Si)$_2$ phase appears and grows when the hot infiltration time exceeds a certain point.
(2) Within 50 min of hot infiltration, the dense Ti(Al,Si)$_3$ phase layer and *L*-(Al, Si, Ce) phase layer are formed outward from the substrate, respectively. When the hot infiltration time is extended, the dense Ti(Al,Si)$_3$ phase, *L*-(Al, Si, Ce) phase + bulk τ_2: Ti(Al,Si)$_2$ phase and *L*-(Al, Si, Ce) phase are obtained from the substrate in sequence.
(3) Several new Ti-Al system alloy phase layers (Ti$_3$Al, TiAl and TiAl$_3$, etc.) are formed between the substrate and Ti(Al,Si)$_3$ during the high-temperature oxidation process, which can further prevent the diffusion of oxygen and cracks to the substrate.
(4) Adding Ce can form Ce-rich quaternary phase (TiAl$_6$Si$_2$Ce$_2$) in *L*-(Al, Si, Ce) alloy layer and suppress the formation of the Al-Si-Ti ternary phase and τ_2: Ti(Al,Si)$_2$ phase. The concentration of Ti, Si and Ce in the TiAl$_6$Si$_2$Ce$_2$ phase is much higher than that of these elements in the surrounding *L*-(Al, Si, Ce) alloy, indicating that the element segregation effect occurs after the addition of Ce.

The hot infiltration method is an economical and efficient method for preparing Al-based coatings for high-temperature-resistant titanium alloy. As the α-β phase transition temperature of titanium alloy is about 900 °C, the hot infiltration temperature cannot exceed 900 °C. The key to high-temperature oxidation resistance is to form a dense and continuous multiphase gradient coating, which is to isolate air and avoid cracks caused by thermal stress between layers. A Ti-Al-Si gradient coating added with Ce can inhibit the formation of a loose τ_2 layer and maintain a dense and continuous multiphase layer structure in the high-temperature oxidation process for a long time. Oxygen atoms cannot diffuse into the substrate, and the Ti-Al-Si gradient coating showed excellent high-temperature oxidation resistance. This work provides a new idea for the preparation and structure design of hot dipping aluminum alloy coatings.

Author Contributions: Data curation, W.H. and Y.T.; Investigation, X.H.; Methodology, Z.L.; Project administration, F.L.; Writing—original draft, Z.W. All authors have read and agreed to the published version of the manuscript.

Funding: This research was funded by the Hunan Provincial Natural Science Foundation of China (No. 2021JJ30672), National Natural Science Foundation of China (No. 52171017), Research initiation project of Xiangtan University and College Students' innovation and entrepreneurship training program of Xiangtan University.

Institutional Review Board Statement: Not applicable.

Informed Consent Statement: Not applicable.

Data Availability Statement: Not applicable.

Acknowledgments: The authors gratefully acknowledge the support provided by Fucheng Yin and Key Laboratory of Materials Design and Preparation Technology of Hunan Province, Xiangtan University, Xiangtan, Hunan.

Conflicts of Interest: The authors declare no conflict of interest.

References

1. Tang, J.M. A review of aerospace materials. *Spacecr. Environ. Eng.* **2013**, *30*, 115–121.
2. Dai, J.J.; Zhu, J.Y.; Chen, C.Z.; Weng, F. High temperature oxidation behavior and research status of modifications on improving high temperature oxidation resistance of titanium alloys and titanium aluminides: A review. *J. Alloy. Compd.* **2016**, *685*, 784–798. [CrossRef]
3. Chen, Z.Y.; Liu, Y.Y.; Jin, Y.F.; Ma, X.Z.; Chai, L.H.; Cui, Y.P. Research on 650 °C high temperature titanium alloy technology for aero-engine. *Aeronaut. Manuf. Technol.* **2019**, *62*, 22–30.
4. Li, F.G.; Yang, L.; Zhou, Y.C. Study Advances of high temperature coating for aeroengine to resist marine atmospheric corrosion. *Therm. Spray Technol.* **2019**, *11*, 1–9.
5. Li, M.S. *High Temperature Corrosion of Metals*; Metallurgical Industry Press: Beijing, China, 2001.
6. Yang, Y.; Kushima, A.; Han, W.Z.; Xin, H.L.; Li, J. Self-healing aluminum oxide during deformation at room temperature. *Nano Lett.* **2018**, *18*, 2492–2497. [CrossRef]
7. Xu, Z.F.; Rong, J.; Yu, X.H.; Meng, K.; Zhan, Z.L.; Wang, X.; Zhang, Y.N. The interface structure of high-temperature oxidation-resistant Aluminum-based coatings on Titanium billet surface. *JOM* **2017**, *69*, 1848–1852. [CrossRef]
8. Christoph, L.; Manfred, P. *Titanium and Titanium Alloys: Fundamentals and Applications*; John Wiley & Sons: Hoboken, NJ, USA, 2003.
9. Xiang, Z.D.; Rose, S.; Datta, P.K. Pack deposition of coherent aluminide coatings on γ-TiAl for enhancing its high temperature oxidation resistance. *Surf. Coat. Technol.* **2002**, *161*, 286–292. [CrossRef]
10. Tsaur, C.C.; Rock, J.C.; Chang, Y.Y. The effect of NaCl deposit and thermal cycle on an aluminide layer coated on 310 stainless steel. *Mater. Chem. Phys.* **2005**, *91*, 330–337. [CrossRef]
11. Wang, Y.S.; Xiong, J.; Yan, J.; Fan, H.Y.; Wang, J. Oxidation resistance and corrosion behavior of hot-dip aluminized coatings on commercial-purity titanium. *Surf. Coat. Technol.* **2011**, *206*, 1277–1282. [CrossRef]
12. Zhang, Z.G.; Peng, Y.P.; Mao, Y.L.; Pang, C.J.; Lu, L.Y. Effect of hot-dip aluminizing on the oxidation resistance of Ti–6Al–4V alloy at high temperatures. *Corros. Sci.* **2012**, *55*, 187–193. [CrossRef]
13. Jeng, S.C. Oxidation behavior and microstructural evolution of hot-dipped aluminum coating on Ti-6Al-4V alloy at 800 °C. *Surf. Coat. Technol.* **2013**, *235*, 867–874. [CrossRef]
14. Wu, Y.; Wang, A.H.; Zhang, Z.; Zheng, R.R.; Xia, H.B.; Wang, Y.N. Laser alloying of Ti–Si compound coating on Ti–6Al–4V alloy for the improvement of bioactivity. *Appl. Surf. Sci.* **2014**, *305*, 16–23. [CrossRef]
15. Sadeq, F.O.; Sharifitabar, M.; Afarani, M.S. Synthesis of Ti–Si–Al coatings on the surface of Ti-6Al-4V alloy via hot dip siliconizing route. *Surf. Coat. Technol.* **2018**, *337*, 349–356. [CrossRef]
16. Dai, J.J.; Zhang, F.Y.; Wang, A.M.; Yu, H.J.; Chen, C.Z. Microstructure and properties of Ti-Al coating and Ti-Al-Si system coatings on Ti-6Al-4V fabricated by laser surface alloying. *Surf. Coat. Technol.* **2017**, *309*, 805–813. [CrossRef]
17. Chen, C.; Feng, X.M.; Shen, Y.F. Oxidation behavior of a high Si content Al-Si composite coating fabricated on Ti-6Al-4V substrate by mechanical alloying method. *J. Alloy. Compd.* **2017**, *701*, 27–36. [CrossRef]
18. Hu, X.Y.; Li, F.G.; Shi, D.M.; Xie, Y.; Li, Z.; Yin, F.C. A design of self-generated Ti–Al–Si gradient coatings on Ti–6Al–4V alloy based on silicon concentration gradient. *J. Alloy. Compd.* **2020**, *830*, 154670. [CrossRef]
19. Rahmel, A.; Schütze, M. Mechanical aspects of the rare-earth effect. *Oxid. Met.* **1992**, *38*, 255–266. [CrossRef]
20. Thanneeru, R.; Patil, S.; Deshpande, S.; Seal, S. Effect of trivalent rare earth dopants in nanocrystalline ceria coatings for high-temperature oxidation resistance. *Acta Mater.* **2007**, *55*, 3457–3466. [CrossRef]
21. Huang, N.; Hu, S.; Xie, G.; Zeng, P.; Ru, Q. Effect of rare earth element cerium on mechanical properties and morphology of TiN coating prepared by arc ion plating. *J. Rare Earth.* **2003**, *21*, 380–383.
22. Li, Z.; Liao, C.L.; Liu, Y.X.; Wang, X.M.; Wu, Y.; Zhao, M.X.; Long, Z.H.; Yin, F.C. 700 °C Isothermal section of the Al-Ti-Si ternary phase diagram. *J. Phase Equilib.* **2014**, *35*, 564–574. [CrossRef]
23. Marder, A.R. The metallurgy of zinc-coated steel. *Prog. Mater. Sci.* **2000**, *45*, 191–271. [CrossRef]
24. Ma, S.M.; Li, N.; Zhang, C.; Wang, X.M. Evolution of intermetallic phases in an Al-Si-Ti alloy during solution treatment. *J. Alloy. Compd.* **2020**, *831*, 154872. [CrossRef]
25. Zhou, W.; Zhao, Y.G.; Li, W.; Qin, Q.D.; Tian, B.; Hu, S.W. Al–Si coating fused by Al + Si powders formed on Ti–6Al–4V alloy and its oxidation resistance. *Mater. Sci. Eng. A* **2006**, *430*, 142–150. [CrossRef]
26. Knaislová, A.; Novák, P.; Kopeček, J.; Průša, F. Properties comparison of Ti-Al-Si alloys produced by various metallurgy methods. *Materials* **2019**, *12*, 3084. [CrossRef] [PubMed]
27. Knaislová, A.; Novák, P.; Linhart, J.; Szurman, I.; Skotnicová, K.; Juřica, J.; Čegan, T. Structure and properties of cast Ti-Al-Si alloys. *Materials* **2021**, *14*, 813. [CrossRef] [PubMed]
28. Knaislová, A.; Novák, P.; Cabibbo, M.; Jaworska, L.; Vojtěch, D. Development of TiAl–Si Alloys—A Review. *Materials* **2021**, *14*, 1030. [CrossRef]

Article

Effect of Cryogenic Treatment on the Microstructure and Wear Resistance of 17Cr2Ni2MoVNb Carburizing Gear Steel

Yongming Yan [1,*], Zixiang Luo [2], Ke Liu [3], Chen Zhang [3], Maoqiu Wang [1] and Xinming Wang [2]

[1] Central Iron & Steel Research Institute Company Limited, Beijing 100081, China; maoqiuwang@hotmail.com
[2] School of Material Science and Engineering, Xiangtan University, Xiangtan 411105, China; 201921001459@smail.xtu.edu.cn (Z.L.); wangxm@xtu.edu.cn (X.W.)
[3] Jianglu Machinery and Electronics Group Co., Ltd., Xiangtan 411100, China; liuke820x@126.com (K.L.); zc719500@163.com (C.Z.)
* Correspondence: yanyongming@nercast.com; Tel.: +86-010-62182728

Abstract: Cryogenic treatment as a process that can effectively improve the performance of steel materials is widely used because of its simplicity and speed. This paper investigates the effects of different low temperature treatments on the microstructure and properties of 17Cr2Ni2MoVNb steel. The low temperature treatment range is divided into cryogenic treatment (CT-80), shallow cryogenic treatment (SCT-150) and deep cryogenic treatment (DCT-196), all with a duration of 1 h. The retained austenite content and the change in carbide volume fraction at 0.2 mm in the carburised layer are studied. The microhardness gradient of the carburised layer, as well as the friction coefficient and wear scar morphology at 0.2 mm, was investigated. The results show that the low temperature treatment is effective in reducing the retained austenite content and increasing the volume fraction of carbide. The lowest retained austenite content and highest carbide volume fraction were obtained for DCT-196 specimens at the same holding time. Due to the further transformation of martensite and the diffuse distribution of carbides, the microhardness and frictional wear properties of DCT-196 are optimal. Therefore, low temperature treatment can change the microstructure of the case layer of 17Cr2Ni2MoVNb steel and effectively improve the mechanical properties of materials.

Keywords: cryogenic treatment; 17Cr2Ni2MoVNb; wear; gear steel; carbide; retained austenite

Citation: Yan, Y.; Luo, Z.; Liu, K.; Zhang, C.; Wang, M.; Wang, X. Effect of Cryogenic Treatment on the Microstructure and Wear Resistance of 17Cr2Ni2MoVNb Carburizing Gear Steel. *Coatings* **2022**, *12*, 281. https://doi.org/10.3390/coatings12020281

Academic Editor: Alessio Lamperti

Received: 17 January 2022
Accepted: 14 February 2022
Published: 21 February 2022

Publisher's Note: MDPI stays neutral with regard to jurisdictional claims in published maps and institutional affiliations.

Copyright: © 2022 by the authors. Licensee MDPI, Basel, Switzerland. This article is an open access article distributed under the terms and conditions of the Creative Commons Attribution (CC BY) license (https://creativecommons.org/licenses/by/4.0/).

1. Introduction

The 17Cr2Ni2MoVNb steel developed from 18CrNiMo7-6 steel undergoes low distortion during carburizing heat treatment. The carburised layer presents a very complex microstructure due to the carbon concentration gradient distribution. It is composed of high carbon tempered martensite, retained austenite and carbides close to the surface, which provide high hardness and good wear resistance. The matrix is composed of low carbon tempered martensite, which supports sufficient toughness and strength. The 17Cr2Ni2MoVNb steel exhibits high strength, toughness, and fatigue properties, making it the potential material for high-strength carburizing steel. However, 17Cr2Ni2MoVNb steel contains a considerable amount of Ni, which enhances the stability of austenite. After carburizing, a substantial amount of carbon and alloying elements is dissolved in the matrix, which significantly reduces the start temperature of martensite transformation (Ms point). A large amount of austenite is retained in the carburized layer after carburizing and air cooling. Reheating and quenching cannot reduce the content of retained austenite, and the hardness cannot be further improved to meet the design requirements. In addition, the retained austenite is heated or transformed under strain during use, causing dimensional changes and redistribution of stress, which may also cause cracks during grinding and finishing. At present, the main ways to reduce the amount of retained austenite are high-temperature tempering and cryogenic treatment [1–4]. Cryogenic treatment has been widely used for tools, bearings, gears, and other workpieces made of high speed-steel, hard alloys and other materials. As an example, the effect of cryogenic treatment on the wear resistance of 20CrNi2Mo steel was studied by Preciado and co-workers [5]. Their results showed that the wear resistance of the carburized sample was, respectively, improved by 17% and 25.5% after cryogenic treatment at −80 and −196 °C compared to conventional heat treatment.

Meanwhile, Kara's team has been systematically studying the application of deep cooling treatment to tools, tool steels and bearing steels [6–9]. Çiçek [6] found that cryogenic treatment with tempering treatment increases the average size of α-phase and forms η-phase in WC-CO tools, thus increasing the microhardness of the tools. Gunes [7] studied the wear behaviour of bearing steel AISI 52100 held at −145 °C for different times. It was shown that 36 h is the optimum holding time. At this holding time, the wear rate and friction coefficient decrease and the hardness reaches its maximum. Kara [8] applied cryogenic treatment to AISI D2 tool steel and ceramic tools. Artificial intelligence methods called artificial neural networks (ANNs) are used to estimate surface roughness based on cutting speed, cutting tool, workpiece, depth of cut, and feed rate. It was shown that the samples subjected to the DCTT-36 process gave the best results in terms of surface roughness and tool wear. The effect of cryogenic treatment on the surface roughness and wear of AB2010 tools was significantly improved. The highest micro and macro hardness values were obtained under the DCT-36 process. The secondary carbide formation of the samples was more uniform in the DCTT-36 process. Kara [9] used the L18 orthogonal experimental table to determine the best surface roughness (Ra) worth control factors for AISI 5140 steel. By using in the cylindrical grinding process of AISI 5140 steel and identified optimum grinding conditions via the Taguchi optimization method.

Current research on 17Cr2Ni2MoVNb steel has mainly focused on the study of the influence of process parameters such as the carburizing, quenching and tempering temperature on the structure and properties of the steel. The effect of the quenching temperature on the microstructure and rolling contact fatigue behaviour of 17Cr2Ni2MoVNb steel was investigated by Zhang and Qu [10,11]. The results showed that failure mode of the sample quenched at 900 °C is delamination, whereas the failure mode after quenching at 1100 °C is pitting corrosion. The effects of the tempering temperature on the tensile properties and rotary bending fatigue behaviour of this steel were also investigated, demonstrating that the tensile strength and fatigue strength were the highest at 180 °C, reaching 1456 and 730 MPa, respectively. At present, however, there are few reports on the effect of cryogenic treatment on the structure and mechanical properties of 17Cr2Ni2MoVNb steel after carburization. In this study, the changes in the microstructure and mechanical properties of 17Cr2Ni2MoVNb steel subjected to cryogenic treatment at three different temperatures are discussed and compared with those of the sample without cryogenic treatment.

2. Experiment

2.1. Materials and Heat Treatment

The experimental materials used in the present study were provided by the Central Iron and Steel Research Institute. The chemical composition of this steel was tabulated in Table 1.

Table 1. Chemical composition of 17Cr2Ni2MoVNb (wt.%).

C	Si	Mn	Cr	Ni	Mo	Nb	V	Fe
0.17	≤0.4	0.77	1.68	1.60	0.29	0.04	0.10	Bal.

Carburizing and tempering are important means of obtaining high hardness and high wear resistance on the surface while ensuring that the matrix has high toughness [12]. Therefore, all samples investigated in this work were first subjected to conventional heat treatment, followed by cryogenic treatment.

Conventional heat treatment consists of normalizing, thermal refining, carburizing, quenching in oil, and tempering. Round steel of diameter of 120 mm was forged into a ring sample with an inner diameter of 70 mm, an outer diameter of 211 mm and a height of 40 mm. The ring samples were normalized, quenched, tempered, and then cut into samples of dimensions 20 mm × 20 mm × 20 mm. The cubic samples were prepared using the process shown in Figure 1. The continuous BH gas carburizing temperature was 930 °C and the total carburizing time was 10 h, then carburizing was followed by high-temperature tempering at 620 °C for 4 h. The samples were held at 800 °C for 1 h, then quenched in oil. After quenching, the low temperature tempering at 150 °C was carried out for 4 h. The carbon content of the sample from the matrix to the surface increased from 0.17% to approximately 1% after carburizing. The increased carbon content leads to the changes in the composition and microstructure of the carburized layer. Cryogenic treatment is employed to eliminate the retained austenite and improve the hardness and wear resistance of the carburized layer [12–16].

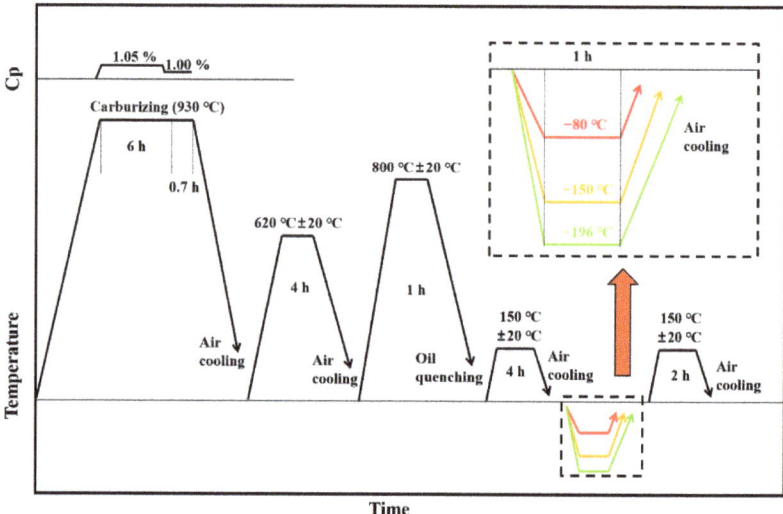

Figure 1. Schematic illustration of the carburizing and heat treatment processes of the experimental steel. Cp denotes the carbon potential during the carburizing process.

Liquid nitrogen was used as the cooling medium. The liquid nitrogen immersion method used in traditional cryogenic treatment easily leads to the formation of the cracks. Therefore, the equipment developed based on the aerosol method, (GL-150 cryogenic treatment experimental furnace produced by Shanghai GaoLe Machinery Equipment Co., Ltd. (Shanghai, China)) shown in Figure 2 was used to avoid the formation of the cracks. The samples for cryogenic treatment were suspended using a wire. The entire experiment was equipped with a real-time temperature sensor for temperature control. The three cryogenic treatment temperatures were designated as −80, −150 and −196 °C as shown in Table 2, and the corresponding samples are denoted as CT-80, SCT-150 and DCT-196, respectively. After each cryogenic heat treatment process, the samples were subjected to low-temperature tempering at 150 °C for 2 h.

Table 2. Process name and description.

Process Name	Process Description
Conventional Heat Treatment (CHT)	-
Cryogenic treatment (CT-80)	Hold at −80 °C for 1 h
Shallow cryogenic treatment (SCT-150)	Hold at −150 °C for 1 h
Deep cryogenic treatment (DCT-196)	Hold at −196 °C for 1 h

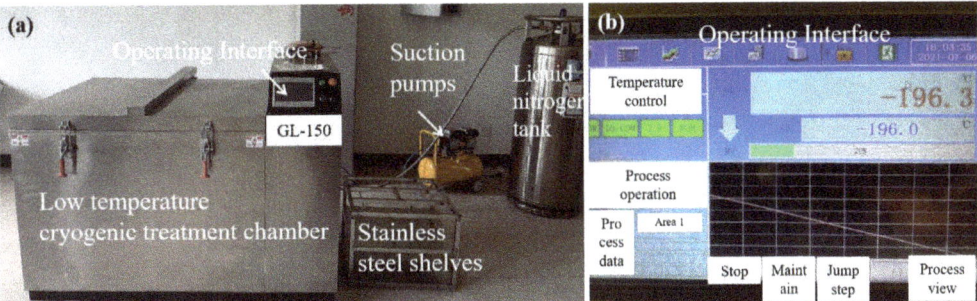

Figure 2. (a) Cryogenic treatment chamber, and (b) cryogenic treatment real time display interface.

2.2. Microstructural Characterization

The scanning technology can provide an important basis for explaining the influence of cryogenic treatment on the wear resistance of materials from the microstructure change [12,14,16]. The samples were cut into 10 mm × 10 mm × 10 mm cubes using an electric spark wire, and then polished with 400, 600 and 800 mesh sandpaper after inlay. Etching was performed using 4% nitric solution for a period of 10 s followed by rinsing with water and drying using a blower. The microstructures of the samples before and after cryogenic treatment were evaluated using a Zeiss EVO MA-10 metallographic microscope (Carl Zeiss, Jena, Germany), scanning electron microscope in bright field (SEM, Carl Zeiss, Jena, Germany)) and energy disperse spectroscopy (EDS, Carl Zeiss, Jena, Germany)).

The content of retained austenite affects the short fatigue crack growth, fatigue performance and wear resistance of carburized steel [5,12]. The content of retained austenite was determined using an X-ray diffractometer (Rigaku, D/MAX-2500/PC (Tokyo, Japan)) with a Cu X-rays source at a power of 18 kW, using a scanning speed for 1°/min. Data were acquired at diffraction angles of 40°–90°.

2.3. Microhardness and Wear Resistance Tests

The microhardness of all the specimens was measured using a Vickers hardness testing machine (HUAYIN, HVT-1000A, Laizhou Huayin, Laizhou, China) with a diamond indenter. The microhardness was measured according to Chinese National Standard GB/T9451-2005 "Determination and Calibration of the depth of the carburizing and quenching hardened layer of steel parts". A load of 9.8 N was applied for a time-period of 10 s. To improve the accuracy, the hardness was determined at five distinct points in different regions of the specimen and the average value was taken. Friction and wear experiments were carried out using a CFT-1 comprehensive tester (Zhongke Kaihua, Lanzhou, China) to evaluate the material surface as shown for Figure 3. The wear mode was reciprocating, the friction surface of the sample was a carburized surface, and the dimensional area is 10 mm × 10 mm. The friction pair was Si_3N_4 ceramic balls, the friction conditions were: 20 °C, dry friction, load of 30 N, reciprocating speed of 500 times/minute, and reciprocating distance of 5 mm. The wear mark depth and wear volume were measured using a probe sensor, and tested after manual zero setting. The measurement method involved taking the upper, middle and lower three points of a wear mark, and the experimental data were acquired three times at one point.

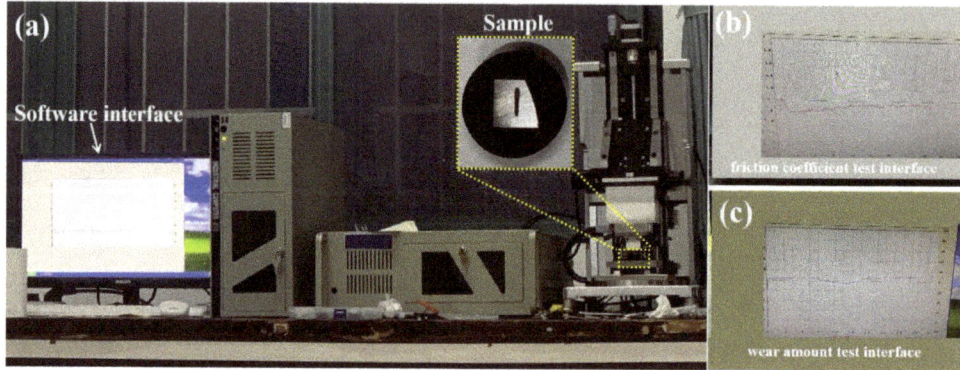

Figure 3. (a) CFT-1 comprehensive material surface property tester, (b) friction coefficient test interface, and (c) wear amount test interface.

3. Results and Discussion

3.1. Effect of Cryogenic Treatment on Microstructure

The changes in the microstructure including changes in the martensite structure, retained austenite and carbides after cryogenic treatment were investigated by SEM and X-ray Diffraction (XRD). Figure 4a,b shows the SEM images of the carburized layer at 0.2 mm from the surface in the four samples, indicating that the microstructure of the 17Cr2Ni2MoVNb steel is mainly composed of plate martensite (gray), retained austenite (concave) and finely dispersed precipitates of carbides (small white grains), the same as that of other carburized steels [5]. As shown in Figure 4a, the sample subjected to the CHT process contained a small amount of carbide precipitations, large areas of carbide accumulation were not detected. During the three cryogenic treatment processes, the amount of carbide precipitated along the grain boundary increased, and the carbide grew to form discontinuous networks, as shown in Figure 4b–d. An EDS point of the local enlargement of Figure 4d was performed to analyse the elemental composition of carbide and matrix, and the results are shown in Table 3. At least 10 regions were analysed for each specimen. The content of C and Cr elements at points P_1 and P_2 is higher than that in the matrix, thus it is speculated that Cr-rich carbide is present in the carburized layer.

The carbide content and the carbide distribution with relative frequency (the number of carbides per area) were correlated with the size of carbides 0.2 mm below the surface using the Image-Pro Plus software (version 6.0) [17]. The images were first processed with Photoshop (PS) to obtain Figure 4e–h and then processed with area statistics of Image-Pro. The calculated data is converted into μm units by scale. The obtained data were processed using origin software [17] using a built-in Gaussian function to fit the histogram data. The final volume fraction and carbide size distribution graph were obtained. The results are shown in Figure 4i–l. The calculated carbide content is 6.15%, 8.3%, 8.7% and 8.9% for the CHT, CT-80, SCT-150 and DCT-196 samples, respectively. The size of the majority of the carbides in the four samples was below 0.6 μm, whereas less carbides with a particle size of over 0.4 μm was detected. The relative frequency of the 0–0.2 μm carbides particles and 1.4–1.6 μm particles in DCT-196 was higher than that in the other samples. It can be seen that the number of carbides increased after cryogenic treatment. As the cryogenic treatment causes the lattice constant of iron to tend to contract, this increases the lattice distortion caused by supersaturated carbon atoms, which enhances the thermodynamic driving force for carbon atom precipitation. As a result, the solubility of carbon atoms decreases. In addition, the diffusion of carbons becomes more difficult at low temperatures and the diffusion distance becomes shorter. As a result, a large number of dispersed ultrafine carbides precipitate on the martensite. Huang and Kelkar used M2 steel as the experimental object and concluded that cryogenic treatment promoted carbide precipitation and also contributed to a more diffuse carbide distribution [18,19].

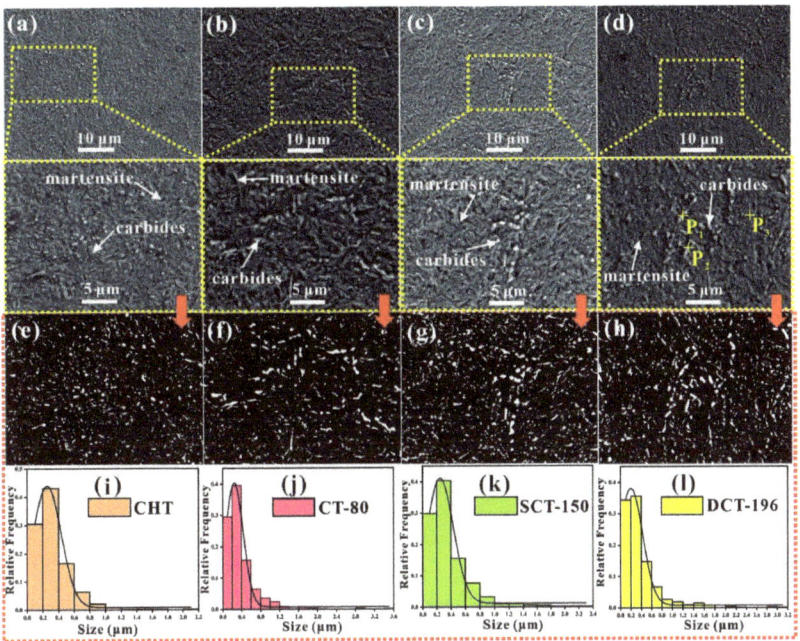

Figure 4. SEM images of the samples at 0.2 mm from the surface: (**a**) CHT, (**b**) CT-80, (**c**) SCT-150, and (**d**) DCT-196; Pictures after PS processing: (**e**) CHT, (**f**) CT-80, (**g**) SCT-150, and (**h**) DCT-196; and histogram of carbide calculation results: (**i**) CHT, (**j**) CT-80, (**k**) SCT-150, and (**l**) DCT-196.

Table 3. EDS results of points P_1, P_2, and P_3.

Sample	Point	at.%						
		C	Ni	Cr	Mn	V	Si	Fe
DCT-196	P1	46.21	0.51	2.68	0.55	-	0.22	49.83
	P2	47.91	0.71	3.85	0.65	0.06	0.24	46.58
	P3	29.60	1.02	1.35	0.25	0.12	0.27	67.37

The SEM microstructure of the core of the sample under different temperatures of cryogenic treatment is shown in Figure 5. As can be seen from Figure 5, the core microstructure of all samples consisted of slatted martensite with carbide. Compared with the CHT samples, the martensite slats of the DCT-196 samples are finer and more nanoscale carbides are dispersed in the martensite slat bundles. The presence of large particles of carbide precipitated on the matrix was also found on CT-80. This phenomenon is attributed to the deflection of carbon atoms and the change in vacancy concentration during the cryogenic process [20,21]. The interstitial carbon atoms and vacancies are transformed into a new grain boundary, and the equilibrium concentration of vacancies is reduced during tempering. During tempering, the vacancy equilibrium concentration rises with temperature, resulting in the addition of new grain boundary to the martensitic structure, leading to the martensite slats refining.

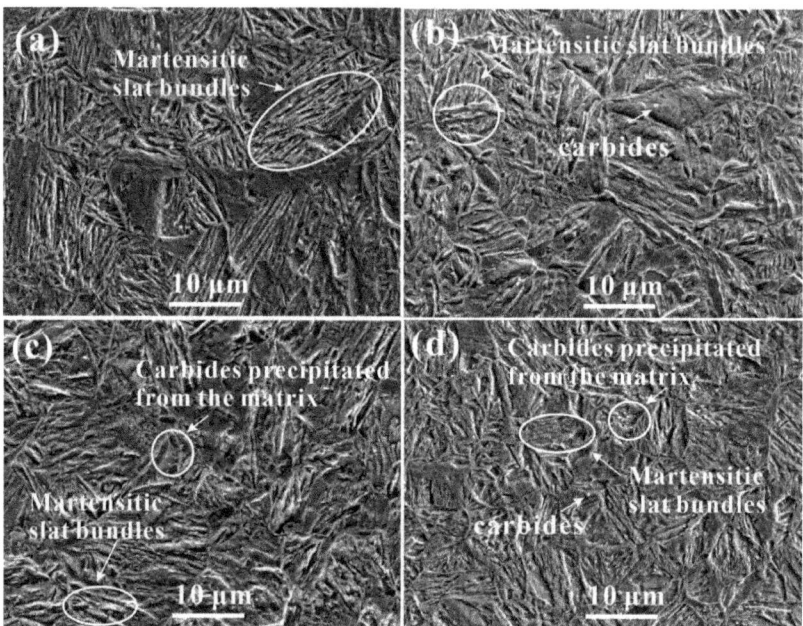

Figure 5. SEM images show the microstructure of the core of (**a**) CHT, (**b**) CT-80, (**c**) SCT-150 and (**d**) DCT-196 specimens.

The XRD patterns of the samples at a distance of 0.2 mm below the surface subjected to the different treatment taken are presented in Figure 6. From Figure 6a, compared with the standard diffraction patterns of matensite and retained austenite, the diffraction patterns of the retained austenite in the tested samples are shifted towards smaller angles, and that of matensite shifted towards smaller angles then towards bigger angles. With the decrease of the cryogenic treatment temperature, the shifting of diffraction patterns of the retained austenite becomes weaken. The standard diffraction patterns of matensite is 06-0696 with cubic structure a = 2.8664, while that of retained austenite is 52-0512 with cubic structure a = 3.618. The lattice parameter was calculated by Jade 6.5 software and the results are shown in Table 4. With the decrease of cryogenic treatment temperature, the lattice constant of martensite gradually increased, and the lattice constant of retained austenite remained basically unchanged after decreasing. The fraction of austenite differed for the various samples, and could be calculated from the XRD results using Equation (1) [22].

$$V_\gamma = \frac{1.4 I_\gamma}{I_\alpha + 1.4 I_\gamma} \qquad (1)$$

where V_γ is the volume fraction of retained austenite, I_γ is the mean integrated intensity of the austenite peaks, including the γ (111), γ (200) and γ (220) peaks; and I_α is the mean integrated intensity of the martensite peaks, including the M/α (110), M/α (200) and M/α (211) peaks. To reduce the error, the lower angle peaks of γ (111) and M/α (110) were neglected [23]. The calculated retained austenite is 18.15%, 12.92%, 10.37% and 9.45% for the CHT, CT-80, SCT-150 and DCT-196 samples, respectively. From the results, it can be seen that cryogenic treatment can significantly reduce the content of retained austenite, but not eliminate it completely. Cryogenic treatment is not only effective in reducing the retained austenite content; it also affects the shape of the retained austenite. After the cryogenic treatment, the retained austenite will change from massive to thin film [24]. The film-like austenite distribution in the steel can effectively relieve the stress concentration in the material during service. This improves the toughness of the material [25].

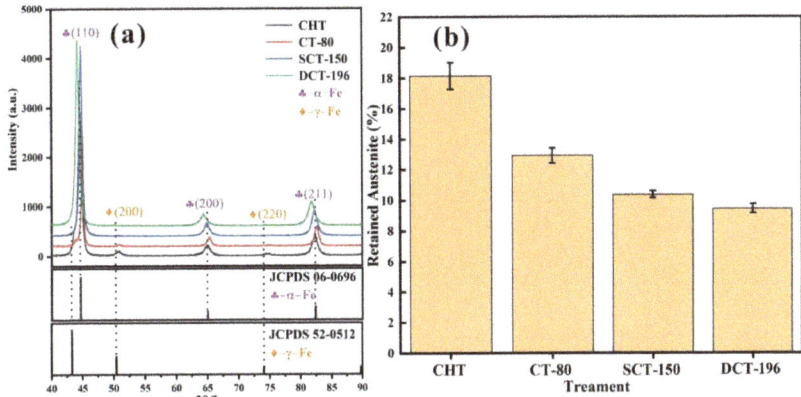

Figure 6. XRD patterns and retained austenite content of 17Cr2Ni2MoVNb steel under different cryogenic treatment processes: (**a**) XRD patterns, and (**b**) retained austenite content.

Table 4. The lattice parameters of the phases in 17Cr2Ni2MoVNb steel under different cryogenic treatment processes.

Sample	Phase	Real Lattice Parameter	Standard Lattice Parameter
CH	Martensite	2.8648	2.8664
	Retained austenite	3.6231	3.618
CT-80	Martensite	2.8745	2.8664
	Retained austenite	3.600	3.618
SCT-150	Martensite	2.8803	2.8664
	Retained austenite	3.600	3.618
DCT-196	Martensite	2.8804	2.8664
	Retained austenite	3.601	3.618

3.2. Effect of Cryogenic Treatment on Mechanical Properties

To reveal the effect of the cryogenic treatment on the mechanical properties of alloy steel, the microhardness and wear resistance were measured. Figure 7 shows the microhardness depth profiles of the carburized alloy steel subjected to CHT, CT-80, SCT-150 and DCT-196. Figure 7 clearly shows that the microhardness of the carburized samples subjected to cryogenic treatment was higher near the surface than that of the sample without cryogenic treatment. The carbon concentration on the surface of the specimen is increased after the carburizing process, followed by high temperature tempering and quenching process. At high temperatures, the carbon element undergoes a redox reaction with the oxygen, which leads to a decrease in the carbon concentration on the surface of the sample and forms an internal oxide layer in the carburized layer [26]. There is a small difference between the depth of the hardened layer (at 550 HV) as cryogenic treatment increased the depth of the hardened layer. Owing to the transformation of retained austenite into martensite and the precipitation of carbide in the carburized layer during the cryogenic treatment, the microhardness of the carburized layer is higher than that of the sample without cryogenic treatment. The DCT-196 sample exhibited the highest microhardness. As the cryogenic treatment temperature decreased, the content of retained austenite close to the surface decreased, and the microhardness increased accordingly. However, as the distance increased and the carbon content in the matrix decreased, the content of retained austenite in the sample decreased, and the change in the hardness became less pronounced, owing to the transformation of retained austenite during the cryogenic treatment.

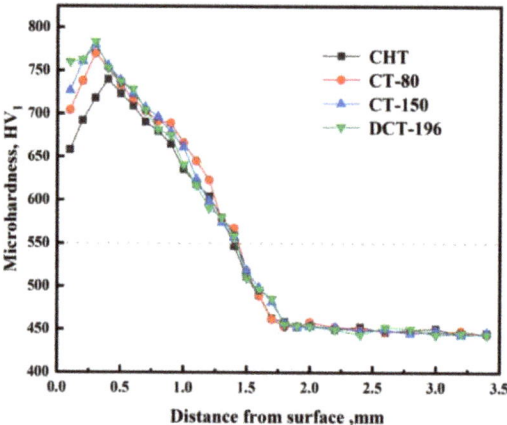

Figure 7. Microhardness of 17Cr2Ni2MoVNb alloys steel under different heat treatments.

Figure 8a shows the relationship between the friction coefficient and time for which 17Cr2Ni2MoVNb steel was subjected to the CHT, CT-80, SCT-150 and DCT-196 processes. At the beginning of the application of friction, the friction coefficients of all the samples increased rapidly. The friction coefficient of the CHT and CT-80 samples increased with time. The friction coefficient of the DCT-196 and SCT-150 samples fluctuated slightly in the first 2 min, and then remained stable at 0.64 and 0.65, respectively. The hardness of the sample is improved due to the reduction of the retained austenite content on the surface after cryogenic treatment. The increase in hardness helps to prevent the frictional substrate from embedding in the substrate and the degree of damage to the substrate is reduced. At the same time, the fine carbides are more uniformly dispersed on the substrate, and the dispersed carbides can effectively impede the dislocation movement of the substrate and increase the wear resistance of the substrate material [27,28].

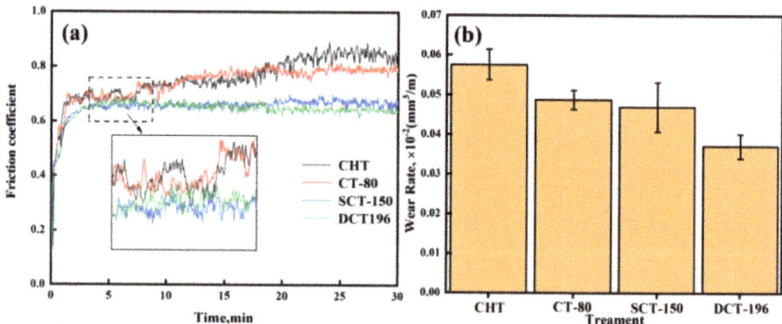

Figure 8. Friction coefficient versus time of 17Cr2Ni2MoVNb alloy steel at different treatment (**a**) and the wear rate after 30 min (**b**).

In addition, the depth of the wear scar and the wear volume were tested, subjected to wear for 30 min. The depths of the wear scar were 22.73, 21.85, 19.01 and 18.65 μm for CHT, CT-80, SCT-150 and DCT-196, respectively. The wear volumes were 0.086, 0.073, 0.070 and 0.055 mm^3 for CHT, CT-80, SCT-150 and DCT-196, respectively. The wear rate was calculated to be 0.0576, 0.0487, 0.0470 and 0.0370 × 10^{-2} mm^3/m for CHT, CT-80, SCT-150 and DCT-196 sample, respectively, as shown in Figure 8b and Table 5. The wear resistance of the samples improved after cryogenic treatment. The wear resistance of the carburized steel improved by 15.37%, 18.47% and 35.68% after cryogenic treatment at −80, −150 and −196 °C, respectively, compared to that of the sample subjected to CHT. The DCT-196 sample had the most stable and lowest friction coefficient, with no significant fluctuation. Here, the wear behaviour of the counterparts to the cryogenic treated samples can be explained by the enhanced amounts of carbides in the latter, which confers good wear resistance compared to the intensive wear experienced by the counterparts. Conversely, this can also be explained by the fact that the matrix of the material has higher hardness, leading to more intensive wear on the counterpart's side.

Table 5. The friction and wear test results of the samples.

Frictional Wear Test Results/Sample	CHT	CT-80	SCT-150	DCT-196
wear scar/μm	22.73	21.85	19.01	18.65
wear volumes/mm^3	0.086	0.073	0.070	0.055
wear rate/×10−2 mm^3/m	0.0576	0.0487	0.0470	0.0370
growth rate	0	15.37%	18.47%	35.68%

Figure 9 shows the SEM images of the wear surfaces of the 17Cr2Ni2MoVNb alloy steel subjected to the CHT, CT-80, SCT-150 and DCT-196 processes. The image of the wear surface shows irregular lumpy areas of different contrast and a larger number of grooves parallel to the wear direction. From EDS analysis, the Si and O contents in the dark irregular block areas were 2.09 at.% and 28.01 at.% in the CHT sample, and 1.65 at.% and 26.59 at.% in the CT-80 sample. The change in the friction coefficient and wear morphology is related to the surface state of the alloy steel and the transformation of the wear mechanism. At the initial stage of sliding friction, micro-convex bodies on the alloy steel surface interact first with the opposite Si_3N_4 ball causing a sharp increase in the friction coefficient. As the friction distance increases, some micro-convex bodies are deformed or even worn away. Some of them will remain on the wear surface, resulting in a high level of friction coefficient and the formation of furrows. As more friction is applied, the wear surface of the alloy steel begins to oxidize to form oxide because of the increase in the surface temperature caused by the accumulated frictional heat. It is generally believed that the oxide film formed on the wear surface acts as a solid lubricant with an anti-friction effect. Therefore, the friction coefficient starts to decrease when a certain area of oxide layer is formed on the wear surface. Under the repeated action of frictional force, the oxide film cracks and partially spalls owing to fatigue, after which oxide growth continues in the exposed area [29–31].

The SEM image of the CHT sample shows a larger number of dark irregularities, whereas that of the SCT-150 has a small number of dark irregularities, and the furrows are clearly defined. Crushing and peeling of the oxide layer were clearly seen in the locally enlarged images in Figure 9a,d. Therefore, the friction and wear behaviour of the samples were classified as abrasive wear and oxidation wear.

Combined with the friction coefficient curve in Figure 8a and the wear trace morphology in Figure 7, a correlation between the friction products and the friction coefficient was observed. When the oxidation and spalling processes are in dynamic equilibrium, the friction coefficient remains stable. The friction coefficient should be reduced if the surface hardness increases. The main reasons are as follows: on one hand, the load-bearing capacity and resistance to deformation of the micro-convex bodies are improved, leading to a decrease in the tendency toward adhesion and consequently in the friction coefficient. On the other hand, the stress per unit area increases (the actual contact area decreases), resulting in a higher local temperature and faster oxidation rate [32].

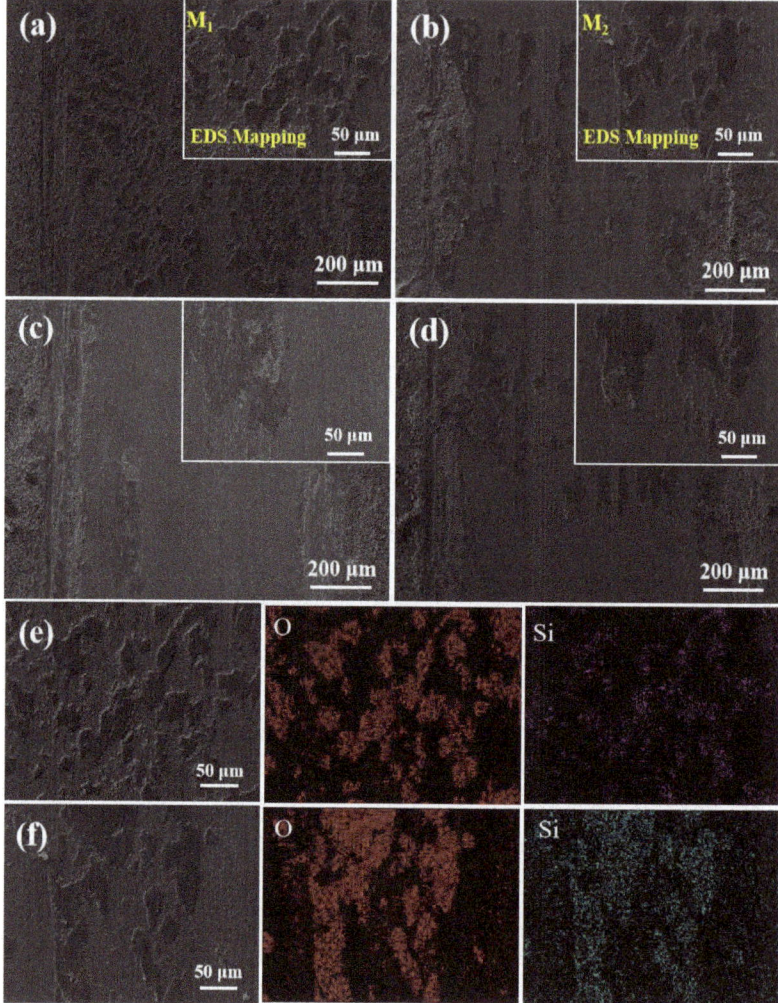

Figure 9. SEM images of the worn-out surfaces of the different samples. (a) CHT, (b) CT-80, (c) SCT-150, and (d) DCT-196. Local magnification of sample with EDS surface scan. (e) CHT, (f) CT-80.

4. Conclusions

The effects of three different temperature (CT-80, SCT-150, DCT-196) cryogenic treatments on the microstructural evolution and mechanical properties of 17Cr2Ni2MoVNb heavy-duty gear steel were investigated. The main conclusions can be drawn as follows:

(1) When the CHT samples were cryogenic treated at three different temperatures, the retained austenite content was significantly reduced. It is reduced from 18.15% for CHT samples to 12.92% for CT-80, 10.37% for SCT-150 and 9.45% for DCT-196, respectively.

(2) The cryogenic treatment can increase the degree of carbide precipitation and dispersion. The volume fraction of carbides was calculated to increase from 6.15% in CHT samples to 8.3% in CT-80, 8.7% in SCT-150 and 8.9% in DCT-196, respectively.

(3) The cryogenic treatment effectively increases the microhardness. The microhardness of each sample increased from 740 HV for CHT samples to 780 HV for CT-80, 780 HV for SCT-150, and 783 HV for DCT-196, respectively.

(4) The wear resistance of the samples improved after the cryogenic treatment. Compared with sample CHT, the wear resistance of CT-80, SCT-150 and DCT-196 increased to 15.37%, 18.47% and 35.68%, respectively. The main wear mechanisms for the samples were abrasive and oxidative wear.

Author Contributions: Conceptualization, Y.Y.; data curation, Z.L.; funding acquisition: Y.Y. and C.Z.; investigation, Y.Y. and Z.L.; methodology, K.L. and C.Z.; project administration, M.W.; validation, K.L.; writing—original draft, Z.L.; writing—review and editing, Y.Y. and X.W. All authors have read and agreed to the published version of the manuscript.

Funding: This work was supported financially by the National Key Research and Development Program under the Grant No. 20T60860B.

Institutional Review Board Statement: Not applicable.

Informed Consent Statement: Not applicable.

Data Availability Statement: Data sharing is not applicable to this article.

Acknowledgments: The authors gratefully acknowledge the support provided by Fucheng Yin and his team members for X-ray diffraction, metallography and electronic microscopy.

Conflicts of Interest: The authors declare no conflict of interest.

References

1. Barron, R.F. Cryogenic treatment of metals to improve wear resistance. *Cryogenics* **1982**, *22*, 409–413. [CrossRef]
2. Zhirafar, S.; Rezaeian, A.; Pugh, M. Effect of cryogenic treatment on the mechanical properties of 4340 steel. *J. Mater. Process. Technol.* **2007**, *186*, 298–303. [CrossRef]
3. SreeramaReddy, T.V.; Sornakumar, T.; VenkataramaReddy, M.; Venkatram, R. Machinability of C45 steel with deep cryogenic treated tungsten carbide cutting tool inserts. *Int. J. Refract. Met. Hard Mater.* **2009**, *27*, 181–185. [CrossRef]
4. Das, D.; Dutta, A.K.; Ray, K.K. Optimization of the duration of cryogenic processing to maximize wear resistance of AISI D2 steel. *Cryogenics* **2009**, *49*, 176–184. [CrossRef]
5. Preciado, M.; Bravo, P.M.; Alegre, J.M. Effect of low temperature tempering prior cryogenic treatment on carburized steels. *J. Mater. Process. Technol.* **2006**, *176*, 41–44. [CrossRef]
6. Çiçek, A.; Kıvak, T.; Ekici, E.; Kara, F.; Uçak, N. Performance of Multilayer Coated and Cryo-Treated Uncoated Tools in Machining of AISI H13 Tool Steel—Part 1: Tungsten Carbide End Mills. *J. Mater. Eng. Perform.* **2021**, *30*, 3436–3445. [CrossRef]
7. Gunes, I.; Cicek, A.; Aslantas, K.; Kara, F. Effect of Deep Cryogenic Treatment on Wear Resistance of AISI 52100 Bearing Steel. *Trans. Indian Inst. Met.* **2014**, *67*, 909–917. [CrossRef]
8. Kara, F.; Karabatak, M.; Ayyıldız, M.; Nas, E. Effect of machinability, microstructure and hardness of deep cryogenic treatment in hard turning of AISI D2 steel with ceramic cutting. *J. Mater. Res. Technol.* **2020**, *9*, 969–983. [CrossRef]
9. Kara, F.; Köklü, U.; Kabasakaloğlu, U. Taguchi optimization of surface roughness in grinding of cryogenically treated AISI 5140 steel. *Mater. Test.* **2020**, *62*, 1041–1047. [CrossRef]
10. Qu, S.-G.; Zhang, Y.-L.; Lai, F.-Q.; Li, X.-Q. Effect of Tempering Temperatures on Tensile Properties and Rotary Bending Fatigue Behaviors of 17Cr2Ni2MoVNb Steel. *Metals* **2018**, *8*, 507. [CrossRef]
11. Zhang, Y.; Qu, S.; Lai, F.; Qin, H.; Huang, L.; Li, X. Effect of Quenching Temperature on Microstructure and Rolling Contact Fatigue Behavior of 17Cr2Ni2MoVNb Steel. *Metals* **2018**, *8*, 735. [CrossRef]
12. Da Silva, V.F.; Canale, L.F.; Spinelli, D.; Bose-Filho, W.W.; Crnkovic, O.R. Influence of retained austenite on short fatigue crack growth and wear resistance of case carburized steel. *J. Mater. Eng. Perform.* **1999**, *8*, 543–548. [CrossRef]
13. Delprete, C. Cryogenic treatment: A bibliographic review. *Med. Equip.* **2008**, *13*, 56–57.
14. Baldissera, P.; Delprete, C. Effects of deep cryogenic treatment on static mechanical properties of 18NiCrMo5 carburized steel. *Mater. Des.* **2009**, *30*, 1435–1440. [CrossRef]
15. Bensely, A.; Senthilkumar, D.; Lal, D.M.; Nagarajan, G.; Rajadurai, A. Effect of cryogenic treatment on tensile behavior of case carburized steel-815M17. *Mater. Charact.* **2007**, *58*, 485–491. [CrossRef]
16. Bensely, A.; Prabhakaran, A.; Lal, D.M.; Nagarajan, G. Enhancing the wear resistance of case carburized steel (En 353) by cryogenic treatment. *Cryogenics* **2005**, *45*, 747–754. [CrossRef]
17. Jovičević-Klug, P.; Jovičević-Klug, M.; Podgornik, B. Effectiveness of deep cryogenic treatment on carbide precipitation. *J. Mater. Res. Technol.* **2020**, *9*, 13014–13026. [CrossRef]
18. Huang, J.Y.; Zhu, Y.T.; Liao, X.Z.; Beyerlein, I.J.; Bourke, M.A.; Mitchell, T.E. Microstructure of cryogenic treated M2 tool steel, Mater. *Sci. Eng. A* **2003**, *339*, 241–244. [CrossRef]
19. Molinari, A.; Pellizzari, M.; Gialanella, S.; Straffelini, G.; Stiasny, K.H. Effect of deep cryogenic treatment on the mechanical properties of tool steels. *J. Mater. Process. Technol.* **2001**, *118*, 350–355. [CrossRef]

20. Li, S.; Deng, L.; Wu, X.; Min, Y.a.; Wang, H. Influence of deep cryogenic treatment on microstructure and evaluation by internal friction of a tool steel. *Cryogenics* **2010**, *50*, 754–758. [CrossRef]
21. Tyshchenko, A.I.; Theisen, W.; Oppenkowski, A.; Siebert, S.; Razumov, O.N.; Skoblik, A.P.; Sirosh, V.A.; Petrov, Y.N.; Gavriljuk, V.G. Low-temperature martensitic transformation and deep cryogenic treatment of a tool steel. *Mater. Sci. Eng. A* **2010**, *527*, 7027–7039. [CrossRef]
22. Sobotova, J.; Jurci, P.; Dlouhy, I. The effect of subzero treatment on microstructure, fracture toughness, and wear resistance of Vanadis 6 tool steel. *Mater. Sci. Eng. A* **2016**, *652*, 192–204. [CrossRef]
23. Putra, W.N.; Pramaditya, P.; Pramuka, P.; Ariati, M. Effect of Sub Zero Treatment on Microstructures, Mechanical Properties, and Dimensional Stability of AISI D2 Cold Work Tool Steel. *Mater. Sci. Forum* **2018**, *929*, 136–141. [CrossRef]
24. Zhang, T.; Hu, J.; Wang, C.; Wang, Y.; Zhang, W.; Di, H.; Xu, W. Effects of deep cryogenic treatment on the microstructure and mechanical properties of an ultrahigh-strength TRIP-aided bainitic steel. *Mater. Charact.* **2021**, *178*, 111247. [CrossRef]
25. Wang, Y.; Yang, Z.; Zhang, F.; Wu, D. Microstructures and mechanical properties of surface and center of carburizing 23Cr2Ni2Si1Mo steel subjected to low-temperature austempering. *Mater. Sci. Eng. A* **2016**, *670*, 166–177. [CrossRef]
26. Song, L.; Gu, X.; Sun, F.; Hu, J. Reduced internal oxidation by a rapid carburizing technology enhanced by pre-oxidation for 18CrNiMo7-6 gear steel. *Vacuum* **2019**, *160*, 210–212. [CrossRef]
27. Gill, S.S.; Singh, J.; Singh, R.; Singh, H. Effect of Cryogenic Treatment on AISI M2 High Speed Steel: Metallurgical and Mechanical Characterization. *J. Mater. Eng. Perform.* **2012**, *21*, 1320–1326. [CrossRef]
28. Tang, L.; Yan, X.; Jiang, Y.; Li, F.; Zhang, H. Corrigendum to "Effect of Magnetic Field Coupled Deep Cryogenic Treatment on Wear Resistance of AISI 4140 Steel". *Adv. Mater. Sci. Eng.* **2021**, *2021*, 9837548. [CrossRef]
29. Hardell, J.; Hernandez, S.; Mozgovoy, S.; Pelcastre, L.; Courbon, C.; Prakash, B. Effect of oxide layers and near surface transformations on friction and wear during tool steel and boron steel interaction at high temperatures. *Wear* **2015**, *330–331*, 223–229. [CrossRef]
30. Das, D.; Ray, K.K.; Dutta, A.K. Influence of temperature of sub-zero treatments on the wear behaviour of die steel. *Wear* **2009**, *267*, 1361–1370. [CrossRef]
31. Yan, X.G.; Li, D.Y. Effects of the sub-zero treatment condition on microstructure, mechanical behavior and wear resistance of W9Mo3Cr4V high speed steel. *Wear* **2013**, *302*, 854–862. [CrossRef]
32. Iakovakis, E.; Roy, M.J.; Gee, M.; Matthews, A. Evaluation of wear mechanisms in additive manufactured carbide-rich tool steels. *Wear* **2020**, *462–463*, 203449. [CrossRef]

Article

Phase Stability and Mechanical Properties of the Monoclinic, Monoclinic-Prime and Tetragonal REMO$_4$ (M = Ta, Nb) from First-Principles Calculations

Wenhui Xiao, Ying Yang, Zhipeng Pi and Fan Zhang *

School of Materials Science and Engineering, Xiangtan University, Xiangtan 411105, China; a18207320474@163.com (W.X.); 18216175931@163.com (Y.Y.); pizhipengmath@163.com (Z.P.)
* Correspondence: zhangfan15@xtu.edu.cn

Abstract: YTaO$_4$ and the relevant modification are considered to be a promising new thermal barrier coating. In this article, phase stability and mechanical properties of the monoclinic (M), monoclinic-prime (M'), and tetragonal (T) REMO$_4$ (M = Ta, Nb) are systematically investigated from first-principles calculations method based on density functional theory (DFT). Our calculations show that M'-RETaO$_4$ is the thermodynamically stable phase at low temperatures, but the stable phase is a monoclinic structure for RENbO$_4$. Moreover, the calculated relative energies between M (or M') and T phases are inversely proportional to the ionic radius of rare earth elements. It means that the phase transformation temperature of M'→T or M→T could decrease along with the increasing ionic radius of RE^{3+}, which is consistent with the experimental results. Besides, our calculations exhibit that adding Nb into the M'-RETaO$_4$ phase could induce phase transformation temperature of M'→M. Elastic coefficient is attained by means of the strain-energy method. According to the Voigt–Reuss–Hill approximation method, bulk modulus, shear modulus, Young's modulus, and Poisson's ratio of T, M, and M' phases are obtained. The B/G criterion proposed by Pugh theory exhibits that T, M, and M' phases are all ductile. The hardness of REMO$_4$ (M = Ta, Nb) phases are predicted based on semi-empirical equations, which is consistent with the experimental data. Finally, the anisotropic mechanical properties of the REMO$_4$ materials have been analyzed. The emerging understanding provides theoretical guidance for the related materials development.

Keywords: phase stability; mechanical properties; modification of YTaO$_4$; lanthanides; first-principles calculations

Citation: Xiao, W.; Yang, Y.; Pi, Z.; Zhang, F. Phase Stability and Mechanical Properties of the Monoclinic, Monoclinic-Prime and Tetragonal REMO$_4$ (M = Ta, Nb) from First-Principles Calculations. *Coatings* 2022, 12, 73. https://doi.org/10.3390/coatings12010073

Received: 21 November 2021
Accepted: 28 December 2021
Published: 8 January 2022

Publisher's Note: MDPI stays neutral with regard to jurisdictional claims in published maps and institutional affiliations.

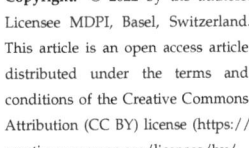

Copyright: © 2022 by the authors. Licensee MDPI, Basel, Switzerland. This article is an open access article distributed under the terms and conditions of the Creative Commons Attribution (CC BY) license (https://creativecommons.org/licenses/by/4.0/).

1. Introduction

The rare-earth tantalate and niobates with the formula REMO$_4$ (M = Ta, Nb) have attracted increasing attention due to their wide application, such as biomedicine, military technology, aerospace, remote sensing, and laser [1]. Moreover, YTaO$_4$ and the relevant modification are extensively investigated and supposed to be promising thermal barrier coatings (TBCs) [2–4] due to high phase stability, good mechanical properties, and thermal conductivity. Because of a ferroelastic toughening mechanism similar to the familiar ZrO$_2$-8 mol%YO$_{1.5}$ (8YSZ) materials, the high-temperature fracture toughness of YTaO$_4$ is very well [5]. It is well known that YTaO$_4$ has three different crystalline structures, such as monoclinic phase (M, space group I2/a), tetragonal phase (T, space group I4$_1$/a), and monoclinic-prime phase (M', space group P2/a). The high-temperature phase transition is a second-order and displacive transformation when the equilibrium tetragonal (T) transited to the monoclinic (M) YTaO$_4$ phase [6]. Although yttrium tantalate has more superior advantages than YSZ, it still has some shortcomings as a new thermal barrier coating. To improve the properties of the yttrium tantalate, doping and modifying are important.

In the periodic table of elements, yttrium and lanthanides belong to the same group of elements and have similar outermost electronic structures, so YTaO$_4$ can be doped with lanthanides to change their properties. Therefore, it is of great significance to systematically study the influence of doping of lanthanide on the mechanical and thermal properties of YTaO$_4$. Up to now, a lot of experimental researches on RETaO$_4$ (RE = La, Nd, Gd, Dy, Yb) have been conducted. The mechanical properties of the M phase are found to be better than M' phase, so it is necessary to stabilize the yttrium tantalate as the M phase below the phase transition temperature. It is studied that the YTaO$_4$ would be stabilized as an M phase when doping 15–30 mol % Nb into YTaO$_4$ materials at 1473 K [7]. It is discovered that the dopant of rare earth elements (Nd, Gd, Dy, Eu, Er, Lu, and Yb) can reduce the thermal conductivity of yttrium tantalate materials [8]. Besides, the mechanical properties and plasticity of RETaO$_4$ (RE = Nd, La, Sm, Gd, Eu, Dy) materials are found to change regularly and become worse and worse with the decrease of atomic radius [9]. In general, yttrium tantalate materials modified by rare earth elements have many advantages, such as great mechanical properties, better thermal stability, and a larger thermal expansion coefficient [10]. Therefore, understanding the doping effects of rare earth elements and Nb on YTaO$_4$ and its phase stability and mechanical properties are significant.

The main purpose of the present work is to systematically investigate the phase stability and mechanical properties of M-, T-, and M'-REMO$_4$ (RE = La, Nd, Gd, Dy, Y; M = Ta, Nb) phases by the first-principles calculation method. Phase stabilities of T-, M-, and M'-RETaO$_4$ or RENbO$_4$ along with the various rare earth elements are studied by comparing their calculated free energies using density functional theory (DFT), and then doping effects of rare earth elements or Nb on YTaO$_4$ are discussed. Elastic stiffness coefficient and elastic flexibility coefficient are attained by means of the strain-energy method. Bulk modulus, Young's modulus, shear modulus, Poisson's ratio, and hardness of T-, M-, and M'-REMO$_4$ (RE = La, Nd, Gd, Dy, Y; M = Ta, Nb) are obtained according to the Voigt–Reuss–Hill approximation method. The B/G criterion proposed by Pugh theory is used to analyze the ductility and brittleness of REMO$_4$ phases. Finally, the anisotropic mechanical properties of the REMO$_4$ materials have been analyzed. It is hoped that the regularity of the YTaO$_4$ materials doped by rare-earth elements or Nb can be determined through first-principles calculations and provide theoretical guidance for the related technological applications.

2. Methods

To theoretically investigate the effect of dopants on the relative stability and mechanical properties of REMO$_4$ (M = Ta, Nb) phases, the first-principles calculations based on density functional theory (DFT) were carried out as implemented in the Vienna Ab-Initio Simulation Package (VASP) [11,12]. The electron-ion interactions were described through projector augmented wave (PAW) [13] and the exchange-correlation functional was constructed by the generalized gradient approximation (GGA) proposed by Perdew–Burke–Ernzerhof (PBE) [14]. The energy cut-off is 550 eV and the converge total energy is less than 1.0 meV/atom. The conjugate gradient method was chosen to relax the structure of atomic positions, cell volumes, and cell shapes. When the residual forces are less than 0.02 eV/Å, the structural relaxations cease. The tetrahedron smearing method with Blochl corrections was used to perform the final self-consistent static calculations [15], and then obtain more accurate energy. For the computation of doping effects of Nb, supercells including 96 atomic sites are used for all structures. A $2 \times 1 \times 2$ supercell is used for the M phase, a $2 \times 2 \times 2$ supercell is used for the M' phase, and a $2 \times 2 \times 1$ supercell is used for the T phase. To model the doping concentration of 25%, 50%, 75%, we choose 4, 8, and 12 Nb atoms and replace the same amount of Ta atoms in supercells, respectively. Taking YTaO$_4$ as an example, the structures of M, M', and T phases are listed in Figure 1.

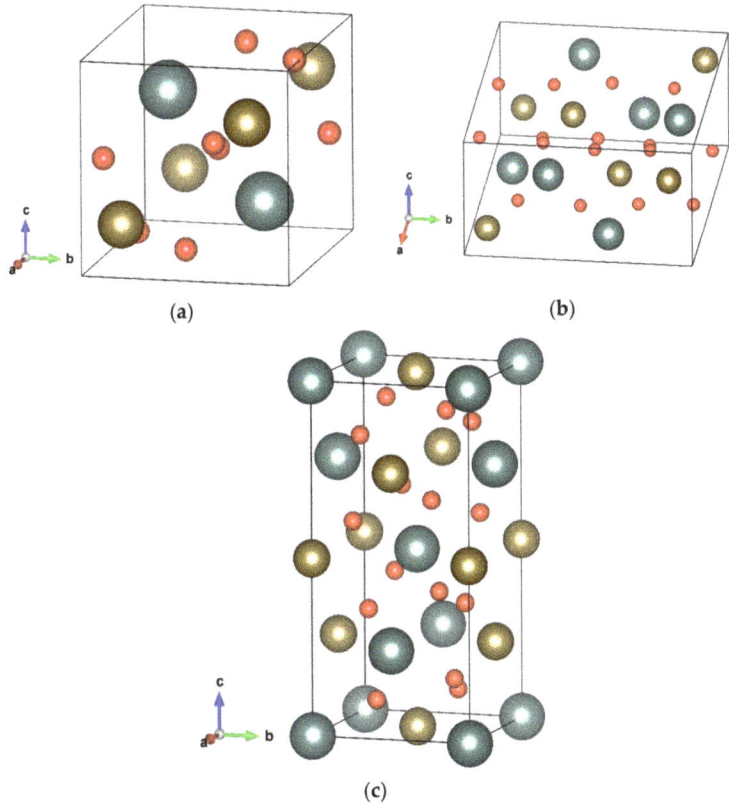

Figure 1. The crystal structures of M, M′, and T phases are listed: The atom of the dark green color is Y; the yellow color is Ta, and the red color is O. (**a**) M′; (**b**) M; (**c**) T.

Elastic Constants

In this work, first-principles calculations are used to calculate the mechanical properties of T-RETa(Nb)O$_4$, M′-RETa(Nb)O$_4$, and M-RETa(Nb)O$_4$ phases). When a very small strain was imposed on the equilibrium crystal, it would exhibit elastic deformation. The strain and stress can be expressed as:

$$\sigma_{ij} = \frac{1}{V_0}\left[\partial E(V,\varepsilon)/\partial \varepsilon_{ij}\right]_{\varepsilon=0} \tag{1}$$

According to the generalized Hooke's law, the elastic constants can be defined as the second derivative of the total energy E (V, ε) to strain,

$$C_{ijkl} = \frac{\partial \sigma_{ij}}{\partial \varepsilon_{kl}} = \frac{1}{V_0}\left[\frac{\partial^2 E(V,\varepsilon)}{\partial \varepsilon_{ij}\partial \varepsilon_{kl}}\right]_{\varepsilon=0} \tag{2}$$

The total energy of a crystal could be expanded using the following Taylor form:

$$E(V,\varepsilon_{ij}) = E(V_0,0) + V_0\sum_{ij}\sigma_{ij}\varepsilon_{ij} + \frac{V_0}{2}\sum_{ijkl}C_{ijkl}\varepsilon_{ij}\varepsilon_{kl} + \dots \tag{3}$$

where E (V$_0$, 0) is the total energy and V$_0$ is the volume of the unstrained system. As shown in the above formula, the strain tensors subscripts (ij, kl) are explained in the Voigt notation scheme (11 = 1, 22 = 2, 33 = 3, 23 = 4, 31 = 5, and 12 = 6) [16]. Before and after the different strains, the total energy variation can be fit by using a multinomial formula. Then we can obtain a secondary coefficient. In this work, 8 distortions to the lattice cell are applied to the lattice cell, and the relaxation in all the strained unit cells was finished when the total energy was converged to less than 1.0 meV/atom.

The Young's modulus (Y), shear modulus (G or μ), bulk modulus (B) and Poisson's ratio (ν) of the polycrystalline crystal were acquired from independent single-crystal elastic constants. In general, the polycrystalline modulus can be approximately assessed by two methods (the Voigt method and the Reuss method) [17,18], and they are expressed as:

$$9B_V = C_{11} + C_{22} + C_{33} + 2(C_{12} + C_{13} + C_{23}) \tag{4}$$

$$\frac{1}{B_R} = (S_{11} + S_{22} + S_{33}) + 2(S_{12} + S_{13} + S_{23}) \tag{5}$$

$$15G_V = (C_{11} + C_{22} + C_{33}) - (C_{12} + C_{13} + C_{23}) + 3(C_{44} + C_{55} + C_{66}) \tag{6}$$

$$15/G_R = 4(S_{11} + S_{22} + S_{33}) - 4(S_{12} + S_{13} + S_{23}) + 3(S_{44} + S_{55} + S_{66}) \tag{7}$$

where the subscripts R and V represent the Reuss and Voigt. The elastic compliance matrices were described as $\{S_{ij}\}$, which is obtained by the inverse matrix of the elastic constant $\{C_{ij}\}^{-1}$. The Voigt–Reuss–Hill approximation [19], which was obtained by the average of Voigt and Reuss bounds, was considered as the best estimation of the polycrystalline elastic modulus. It was indicated as:

$$B_H = (B_V + B_R)/2 \quad G_H = (G_V + G_R)/2 \tag{8}$$

In addition, the Poisson's ratio and the polycrystalline elastic modulus can be obtained using the following relationship:

$$Y_H = \frac{9B_H G_H}{3B_H + G_H} \quad \nu_H = \frac{3B_H - 2G_H}{2(3B_H + G_H)} \tag{9}$$

3. Results

3.1. Structural Properties and Thermodynamic Properties

In the present work, structure relaxations of T, M, and M'-REMO$_4$ (RE = Y, Dy, Gd, Nd, La; M = Ta, Nb) phases were performed. The crystal structures of M-, M'-RENbO$_4$, and RETaO$_4$ phases both belong to the monoclinic crystal structure, and the T phase is the tetragonal crystal. Tables 1 and 2 list the calculated information of the crystal lattice at 0 K and the experimental data [20,21]. Our calculated results are consistently consistent with the experimental values. Both the calculations and experiments show that the small rare earth atom in RETaO$_4$ or RENbO$_4$ phases have small volumes. Besides, the β angle of the M phase and M' phase also gradually decrease with the decrease of the atomic radius of RE^{3+}.

Figure 2a-j show the total energies for T, M, and M'-REMO$_4$ (RE = Y, Dy, Gd, Nd, La; M = Ta, Nb) phases, which are changed with a function of volume at 0 K. The equation of state (EOS) is used to fit the energy-volume. As we know, there are three crystalline structures in RETaO$_4$ materials, and they are monoclinic phase (M, space group I2/a), tetragonal phase (T, space group I41/a), and monoclinic-prime phase (M', space group P2/a). At the high temperature, the stable phase is the T-RETaO$_4$, and it can transform to the M phase through a displacive transformation of T→M. However, the true equilibrium phase at low temperature is the M' phase, which only can be obtained by means of synthesizing below the temperature of T→M transformation. Therefore, the M' phase is the low-temperature phase of the RETaO$_4$ materials. As shown in Figure 2a–j, our calculated results show that the M phase and T phase are both less stable than the M' phase. It implies that the M phase is metastable and the M' phase is stable at low temperatures. This is consistent with the experimental results [22]. For RENbO$_4$ phases, only low-temperature M and high-temperature T phases are existent in the literature. As for comparisons, the M'-RENbO$_4$ structures are also calculated in this work. In Figure 2a–j, our calculations exhibit that the total energy of the M-RENbO$_4$ phase is larger than that of the M' phase, so the M phase is a true equilibrium phase at low temperature. This is consistent with the experimental results [22] that M'-RENbO$_4$ crystalline structures do not exist in the RENbO$_4$ phases.

Table 1. Calculated lattice parameters (Å) of M-, M'-, and T-RETaO$_4$ phases along with the experimental data.

Phase	Abbr.	Group	a	b	c	β	Remark
YTaO$_4$	T	I4$_1$/a	5.23	5.23	11.06		cal
	M'	P2/a	5.15	5.53	5.34	96.40°	cal
			5.26	5.43	5.08	96.08°	exp [20]
	M	I2/a	5.362	11.071	5.093	95.58°	cal
			5.24	10.89	5.06	95.31°	exp [21]
DyTaO$_4$	T	I4$_1$/a	5.24	5.24	11.06		cal
	M'	P2/a	5.34	5.52	5.15	96.58°	cal
		P2/a	5.32	5.48	5.14	96.52°	exp [20]
	M	I2/a	5.36	11.07	5.10	95.51°	cal
		I2/a	5.35	10.97	5.06	95.6°	exp [21]
GdTaO$_4$	T	I4$_1$/a	5.27	5.27	11.17		cal
	M'	P2/a	5.38	5.55	5.19	96.75°	cal
		P2/a	5.36	5.52	5.17	96.66°	exp [20]
	M	I2/a	5.41	11.15	5.11	95.53°	cal
		I2/a	5.41	11.07	5.08	95.6°	exp [21]
NdTaO$_4$	T	I4$_1$/a	5.37	5.37	11.48		cal
	M'	P2/a	5.47	5.66	5.28	96.83°	cal
		P2/a	5.43	5.60	5.24	96.77°	exp [20]
	M	I2/a	5.55	11.40	5.17	95.47°	cal
		I2/a	5.51	11.23	5.11	95.7°	exp [21]
LaTaO$_4$	T	I4$_1$/a	5.44	5.44	11.69		cal
	M'	P2/a	5.52	5.77	5.34	96.75°	cal
	M	I2/a	5.65	11.57	5.19	95.63°	cal
	nM	P2$_1$/c	7.77	5.59	7.86	101.13°	cal
		P2$_1$/c	7.76	5.58	7.81	101.53°	exp [20]

Table 2. Calculated lattice parameters (Å) of M-, M'-, and T-RENbO$_4$ phases along with the experimental data.

Phase	Abbr.	Group	a	b	c	β	Remark
YNbO$_4$	T	I4$_1$/a	5.25	5.25	11.08		cal
	M'	P2/a	5.11	5.45	5.29	96.44°	cal
	M	I2/a	5.31	10.97	5.07	94.42°	cal
		I2/a	5.29	10.94	5.07	94.32°	exp [21]
DyNbO$_4$	T	I4$_1$/a	5.25	5.25	11.11		cal
	M'	P2/a	5.15	5.49	5.37	95.77°	cal
	M	I2/a	5.34	11.11	5.15	93.89°	cal
		I2/a	5.32	11.00	5.07	94.34°	exp [21]
GdNbO$_4$	T	I4$_1$/a	5.29	5.29	11.22		cal
	M'	P2/a	5.18	5.53	5.41	95.85°	cal
	M	I2/a	5.38	11.21	5.18	93.74°	cal
		I2/a	5.37	11.09	5.11	94.37°	exp [21]
NdNbO$_4$	T	I4$_1$/a	5.38	5.38	11.53		cal
	M'	P2/a	5.27	5.66	5.50	95.98°	cal
	M	I2/a	5.47	11.51	5.29	92.42°	cal
		I2/a	5.47	11.28	5.14	94.32°	exp [21]
LaNbO$_4$	T	I4$_1$/a	5.45	5.45	11.74		cal
	M'	P2/a	5.33	5.76	5.55	95.86°	cal
	M	I2/a	5.57	11.53	5.20	94.40°	exp [21]

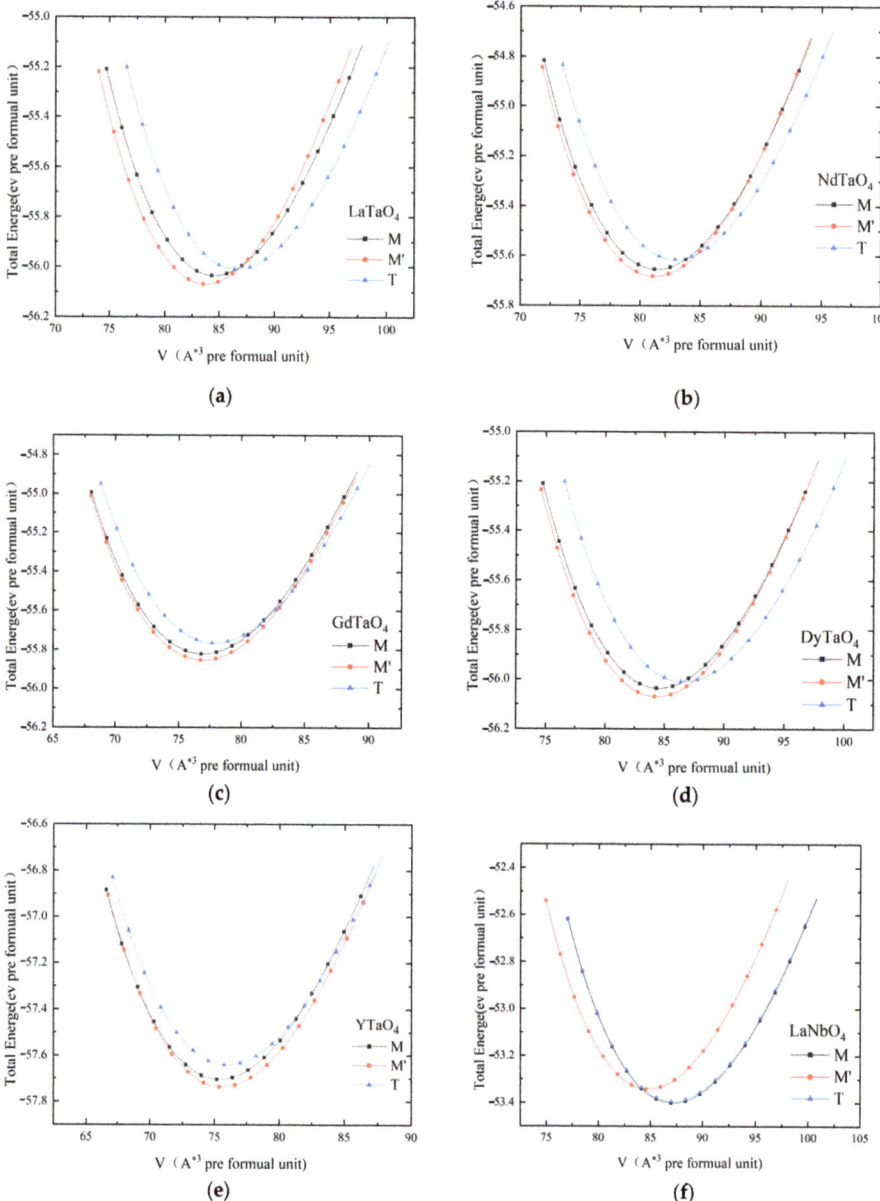

Figure 2. *Cont.*

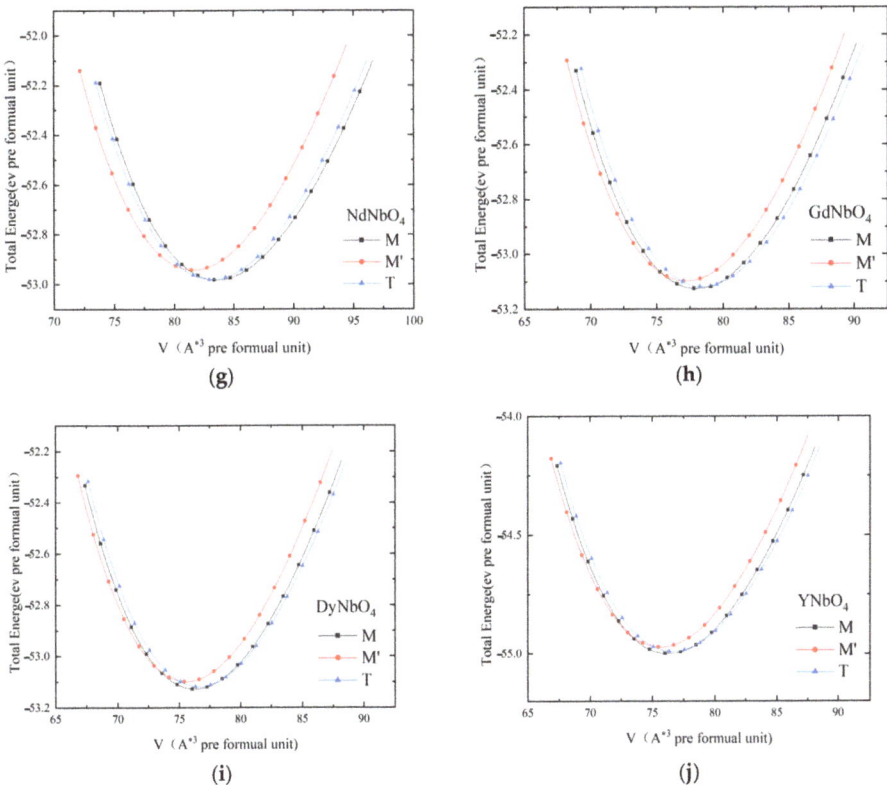

Figure 2. (a–j). Calculated total energies as a function of volume of T- and M-RENbO$_4$ phases and M'-RENbO$_4$ phase: (**a**) LaTaO$_4$; (**b**) NdTaO$_4$; (**c**) GdTaO$_4$; (**d**) DyTaO$_4$; (**e**) YTaO$_4$; (**f**) LaNbO$_4$; (**g**) NdNbO$_4$; (**h**) GdNbO$_4$; (**i**) DyNbO$_4$; (**j**) YNbO$_4$.

Figure 3 presents the relative energies ($\Delta E_{M'\rightarrow T \text{ or } M\rightarrow T}$) of M'- or M-REMO$_4$ with respect to that of T-REMO$_4$ for different rare-earth atoms of Y, Dy, Gd, Nd, and La. It is obvious that the relative energies are inversely proportional to the ionic radius of RE^{3+}. The Gibbs free energy difference of M'→T or M→T phase transformation $\Delta G_{M'\rightarrow T \text{ or } M\rightarrow T}$ can be expressed as $\Delta H_{M'\rightarrow T \text{ or } M\rightarrow T}$ -$\Delta S_{M'\rightarrow T \text{ or } M\rightarrow T}$*T. If the differences in the enthalpy ΔH ($\approx \Delta E$) and entropy ΔS are assumed to be substantially unchanged 4 [23] and ΔS is supposed to be similar for the different rare-earth dopants, phase transformation temperature of M'→T or M→T may increase with the decreasing ionic radius of RE^{3+} for Y, Dy, Gd, Nd, and La. This is consistent with the measured results using a high-temperature X-ray diffractometer by Stubičan [24]. Figure 3 presents the comparison of the experimental transformation temperature and our calculated relative energies, which indicates that the relative energies and transformation temperature decrease with the increase of the rare earth ionic radius.

As M' and M are the stable phases at the low temperature for the RETaO$_4$ and RENbO$_4$ structures, respectively, adding the Nb element into the M'-RETaO$_4$ phase should induce phase transformation of M'→M at the appropriate compositions. Based on the first-principles calculations, phase transformation of M'→M induced by the dopant of Nb is studied in this work. Figure 4 presents our calculated relative energies ($\Delta E_{M'\rightarrow M}$) of M'-RETa$_xNb_{1-x}O_4$ (x = 0.25, 0.5, 0.75) with respect to that of M-RETa$_x$Nb$_{1-x}$O$_4$ (x = 0.25, 0.5, 0.75) for different rare-earth atoms of Y, Dy, Gd, Nd, and La. When $\Delta E_{M'\rightarrow M}$ > 0, this means that M' is thermodynamically stable. Conversely, when $\Delta E_{M'\rightarrow M}$ < 0, it implies that

M is thermodynamically more stable than M′. In Figure 4, the M′-RETa$_x$Nb$_{1-x}$O$_4$ phase will transform into the M-RETa$_x$Nb$_{1-x}$O$_4$ phase when the composition of the dopant Nb is about 0.5. It is worth mentioning that our calculated results are related to the phase transformation at 0 K, and transformation composition may decrease at the high temperature.

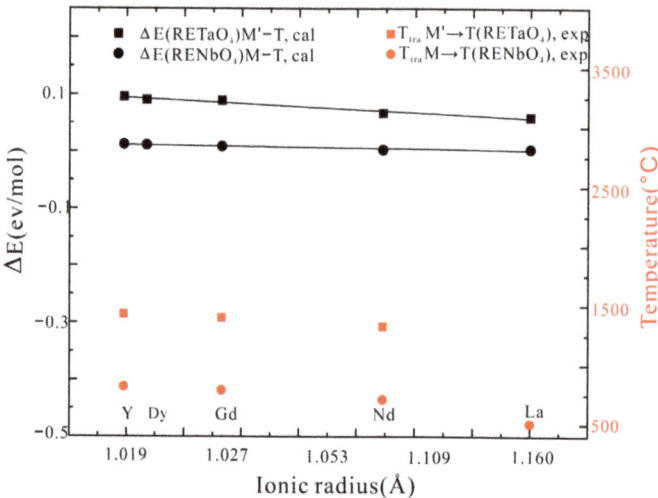

Figure 3. Comparison of the experimental transformation temperature and our calculated relative energies for different rare-earth atoms of Y, Dy, Gd, and Nd.

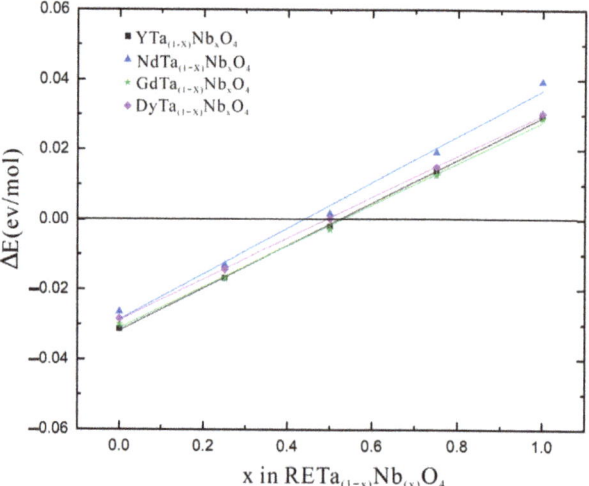

Figure 4. Calculated relative energies ($\Delta E_{M'\rightarrow M}$) of M′-RETa$_xNb_{1-x}O_4$ (x = 0.25, 0.5, 0.75) with respect to that of M-RETa$_x$Nb$_{1-x}$O$_4$ (x = 0.25, 0.5, 0.75) for different rare earth atoms of Y, Dy, Gd, Nd.

3.2. Mechanical Properties

The calculated elastic constants are listed in Tables 3–5. Because the T-YTaO$_4$ phase is tetragonal, C_{11}, C_{12}, C_{13}, C_{33}, C_{44}, and C_{66} can be determined through six deformation modes [25]. As M′-YTaO$_4$ and M-YTaO$_4$ phases are both monoclinic, the thirteen independent elastic constants can be obtained by applying thirteen distortions [26]. The total energies are varied before and after a set of different strains (±1%, ±2%, ±3%, and ±4%), and the elastic constants were calculated by the quadratic coefficients. For stable structures, the elastic constants need to meet the mechanical stability criterion [27]. The monoclinic system criteria are $C_{11} > 0$, $C_{22} > 0$, $C_{33} > 0$, $C_{44} > 0$, $C_{55} > 0$, $C_{66} > 0$, $C_{44}C_{66} - 2C_{46} > 0$, $C_{11} + C_{22} + C_{33} + 2(C_{12} + C_{13} + C_{23}) > 0$, $C_{22} + C_{33} - 2C_{23} > 0$. The tetragonal system criteria are $C_{11} > 0$, $C_{33} > 0$, $C_{44} > 0$, $C_{66} > 0$, $C_{11}\text{-}C_{12} > 0$, $C_{11} + C_{33} - 2C_{13} > 0$, $2C_{11} + C_{33} + 2C_{12} + 4C_{13} > 0$. All RETaO$_4$ (RE = La, Nd, Gd, Dy) and RENbO$_4$ (RE = La, Nd, Gd, Dy) materials meet the criterion of mechanical stability, and the structures are stable.

Table 3. Elastic constants C_{ij} for the M phases of RETaO$_4$ and RENbO$_4$. All quantities are in GPa.

M-RETaO$_4$	C_{11}	C_{22}	C_{33}	C_{44}	C_{55}	C_{66}	C_{12}	C_{13}	C_{23}	C_{16}	C_{26}	C_{36}	C_{45}
La	217.07	168.49	230.98	43.28	55.86	56.26	61.27	106.50	90.21	25.28	6.27	−13.88	−1.14
Nd	229.36	193.08	255.53	50.63	62.02	67.21	63.53	116.64	98.38	21.92	1.88	−19.63	−2.09
Gd	249.50	212.45	270.47	58.12	61.57	83.45	70.90	131.87	95.84	13.11	−2.83	−20.28	−4.73
Dy	255.11	220.50	275.98	57.91	60.10	88.91	74.01	136.94	94.82	10.69	−4.04	−19.74	−4.96
Y	257.02	221.67	273.37	56.72	57.56	87.77	74.78	135.87	93.11	8.91	−4.74	−19.08	−5.23
M-RENbO$_4$	C_{11}	C_{22}	C_{33}	C_{44}	C_{55}	C_{66}	C_{12}	C_{13}	C_{23}	C_{16}	C_{26}	C_{36}	C_{45}
La	171.39	154.27	195.03	30.25	34.82	33.65	63.13	100.72	75.39	34.69	4.56	−23.70	3.09
Nd	157.05	169.74	224.90	29.37	43.43	47.48	49.18	117.27	84.96	45.25	8.44	−23.51	8.18
Gd	195.53	194.15	245.42	42.11	49.75	69.42	64.69	134.61	83.74	29.95	0.79	−22.30	2.39
Dy	205.94	201.75	249.32	45.85	49.62	75.96	68.74	140.48	81.48	25.77	−1.20	−21.93	0.68
Y	202.58	192.94	248.02	44.99	47.80	75.01	65.47	135.69	76.74	25.19	−1.17	−20.13	0.41

Table 4. Elastic constants C_{ij} for the M' phases of RETaO$_4$ and RENbO$_4$. All quantities are in GPa.

M'-RETaO$_4$	C_{11}	C_{22}	C_{33}	C_{44}	C_{55}	C_{66}	C_{12}	C_{13}	C_{23}	C_{16}	C_{26}	C_{36}	C_{45}
La	238.64	162.73	253.65	52.01	60.31	59.34	86.70	113.51	87.66	16.34	−13.07	9.78	−10.43
Nd	266.93	178.92	271.78	62.93	62.67	66.14	91.90	118.95	90.78	15.35	−14.16	8.47	−9.67
Gd	288.41	176.27	296.06	66.70	59.66	71.33	88.50	119.39	90.73	14.69	−15.78	11.01	−7.09
Dy	296.31	173.88	304.01	67.04	56.06	71.97	84.00	118.53	89.13	14.15	−17.72	11.54	−6.91
Y	290.40	141.65	298.91	64.01	53.08	71.22	74.32	113.46	75.89	11.68	−20.59	10.52	−8.10
M'-RENbO$_4$	C_{11}	C_{22}	C_{33}	C_{44}	C_{55}	C_{66}	C_{12}	C_{13}	C_{23}	C_{16}	C_{26}	C_{36}	C_{45}
La	235.54	155.44	220.45	46.37	53.15	47.89	84.43	111.03	84.45	16.31	−6.62	−7.09	−10.34
Nd	258.48	181.92	240.56	53.68	57.23	50.10	91.42	113.48	88.80	17.39	−7.71	−7.47	−9.32
Gd	280.10	189.47	259.83	57.03	57.10	53.72	88.91	110.44	92.54	18.43	−2.49	−3.91	−4.55
Dy	292.68	188.65	271.23	57.82	55.63	55.05	88.54	112.08	96.52	18.88	−0.28	−1.63	−2.45
Y	288.33	181.77	271.43	56.05	53.58	54.05	84.43	110.17	94.51	17.74	−1.31	−1.75	−2.21

Table 5. Elastic constants C_{ij} for the T phases of RETaO$_4$ and RENbO$_4$. All quantities are in GPa.

T-RETaO$_4$	C_{12}	C_{11}	C_{33}	C_{44}	C_{66}	C_{13}	C_{16}
La	124.89	163.70	151.93	29.66	27.52	71.67	113.03
Nd	122.49	194.92	172.31	31.22	28.68	73.87	147.44
Gd	121.73	228.82	192.05	29.00	13.41	74.49	93.62
Dy	120.51	241.46	201.29	27.93	11.56	74.64	148.69
Y	118.20	237.39	198.37	26.25	12.87	71.61	69.08
T-RENbO$_4$	C_{12}	C_{11}	C_{33}	C_{44}	C_{66}	C_{13}	C_{16}
La	112.79	169.36	151.35	31.98	34.65	72.74	128.76
Nd	114.62	196.62	170.36	34.74	29.99	75.69	138.65
Gd	115.24	226.54	190.83	33.89	14.65	77.27	159.76
Dy	116.10	239.29	198.93	33.28	9.67	78.08	164.16
Y	111.40	233.13	197.32	31.69	9.17	75.04	146.88

'The polycrystalline elastic mechanical properties, such as shear modulus (G or μ), bulk modulus (B), Young's modulus (Y), and Poisson's ratio (ν) could be obtained through the Voigt and Reuss methods according to the calculated elastic constants. Using energy considerations, Hill [20] certificated the elastic moduli of the Voigt and Reuss methods are the upper and lower limits of polycrystalline constants. The practical elastic modulus can be estimated by the arithmetic means of these extremes. Generally, the bulk modulus is a measure of resistance to volume change by applied pressure. As seen from Figure 5 and Tables 6–8, the calculated shear modulus and bulk modulus of M-, M′-, T-RETaO$_4$, and RENbO$_4$ (RE = Y, Dy, Gd, Nd, La) is decreased with the increase of the rare-earth atoms, which indicate that the resistance to volume change through applied pressure is eventually lowered. Moreover, the calculated bulk modulus of rare-earth tantalate is regularly larger than and rare-earth niobates. The calculated shear modulus shows a similar trend, which means that the resistance to reversible deformations upon shear stress for RETaO$_4$ and RENbO$_4$ (RE = Y, Dy, Gd, Nd, La) is decreased with the increase of the rare-earth atoms. The ratio between bulk modulus and shear modulus, proposed by Pugh theory [28], can be used to empirically predict the brittleness and ductility of materials. A low B/G ratio is associated with brittleness, and a high value indicates its ductile nature. The empirically critical value which distinguishes ductile and brittle materials is around 1.75. In the present work, the calculated B/G of REMO$_4$ in Figure 6 is larger than 1.75, which means that all REMO$_4$ materials are ductile.

Table 6. Bulk modulus (GPa), shear modulus (GPa), B/G, Young modulus (GPa), Poisson ratio, and Vickers-hardness (Kg·N) for T-REMO$_4$ (M = Ta,Nb) phases.

T-RETaO$_4$	B	G	B/G	E	ν	H$_v$	Remake
La	111.33	29.97	3.71	82.50	0.38	278.11	cal
Nd	120.97	36.28	3.33	98.95	0.36	319.77	cal
Gd	130.55	33.95	3.84	93.74	0.38	270.37	cal
Dy	134.21	33.87	3.96	93.71	0.38	261.17	cal
Y	131.14	33.86	3.87	93.52	0.38	268.04	cal
	128.9	52.7	2.45	139.1	0.32	-	exp [3]
T-RENbO$_4$	B	G	B/G	E	ν	H$_v$	Remark
La	110.57	34.52	3.20	93.80	0.36	333.25	cal
Nd	120.38	38.68	3.11	104.82	0.35	347.78	cal
Gd	130.04	36.34	3.58	99.72	0.37	295.08	cal
Dy	134.24	34.04	3.94	96.69	0.38	262.75	cal
Y	130.41	33.26	3.92	91.97	0.38	263.96	cal

Table 7. Bulk modulus (GPa), shear modulus (GPa), B/G, Young modulus (GPa), Poisson ratio, and Vickers-hardness (Kg·N) for M′-REMO$_4$ (M = Ta,Nb) phases.

M′-RETaO$_4$	B	G	B/G	E	ν	H$_v$	Remark
La	132.73	56.96	2.33	149.48	0.31	481.9	cal
Nd	142.66	64.43	2.21	168.01	0.30	515.73	cal
Gd	144.97	68.23	2.12	176.93	0.30	542.89	cal
Dy	144.00	68.59	2.10	177.04	0.30	550.51	cal
Y	128.67	65.20	2.19	167.34	0.28	588.38	cal
	132.7	66.1	2.01	170.2	0.29	-	exp [3]
M′-RENbO$_4$	B	G	B/G	E	ν	H$_v$	
La	126.58	50.02	2.53	132.60	0.33	434.37	cal
Nd	138.32	56.44	2.45	149.05	0.32	454.79	cal
Gd	143.03	61.34	2.33	161.00	0.31	484.34	cal
Dy	145.98	62.52	2.33	164.14	0.31	484.391	cal
Y	142.59	61.29	2.33	160.75	0.31	485.32	cal

Table 8. Bulk modulus (GPa), shear modulus (GPa), B/G, Young modulus (GPa), Poisson ratio, and Vickers-hardness (Kg·N) for M-REMO$_4$ (M = Ta,Nb) phases.

M-RETaO$_4$	B	G	B/G	E	ν	H$_v$	Remark
La	122.08	52.80	2.31	138.45	0.32	483.34	cal
Nd	134.00	60.47	2.22	157.69	0.31	512.68	cal
	-	-	-	-	-	641	exp [8]
Gd	145.03	67.75	2.14	175.87	0.31	537.98	cal
	-	-	-	-	-	610	exp [8]
Dy	148.88	69.15	2.15	179.61	0.31	535.20	cal
	-	-	-	-	-	534	exp [8]
Y	148.68	68.32	2.18	177.74	0.30	528.35	cal
	183.7	63.2	2.91	170.1	0.34	378	exp [3]
M-RENbO$_4$	B	G	B/G	E	ν	H$_v$	Remark
La	106.00	30.05	3.53	82.38	0.37	286.18	cal
Nd	107.85	39.70	2.72	106.08	0.331	393.53	cal
Gd	129.36	50.99	2.54	135.21	0.33	433.19	cal
Dy	134.53	54.51	2.47	144.08	0.32	449.5	cal
Y	129.95	54.12	2.4	142.57	0.32	463.94	cal

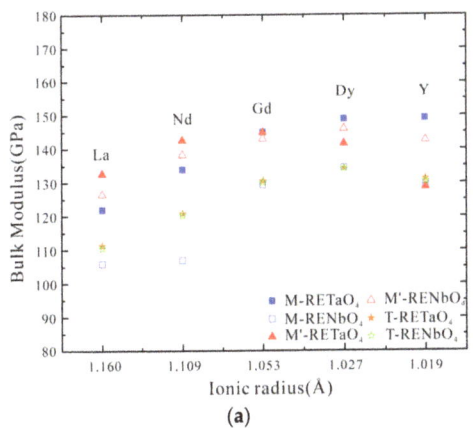
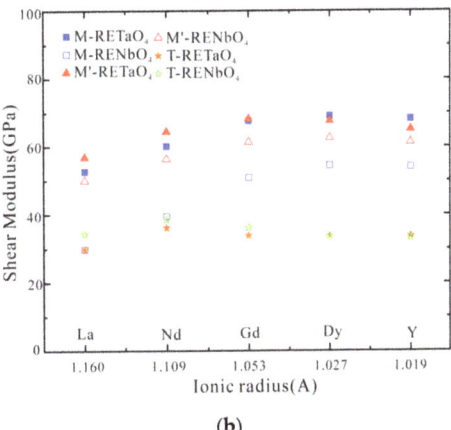

Figure 5. Calculated bulk modulus and shear modulus of M-, M'-, T-RETaO$_4$, and RENbO$_4$ along with the ionic radius of RE^{3+}: (**a**) bulk modulus; (**b**) shear modulus.

Young's modulus E can be used to estimate the stiffness of materials. The calculations of M-, M'-, T-RETaO$_4$ or RENbO$_4$ in Figure 7 and Tables 6–8 suggest that Young's modulus is decreased with an increase of the rare-earth atoms, which means that M-, M'-, and T-YTaO$_4$ are the stiffest, and then followed by DyTaO$_4$, GdTaO$_4$, NdTaO$_4$, and LaTaO$_4$. Moreover, our calculated results indicate that RETaO$_4$ is stiffer than RENbO$_4$. Poisson's ratio (ν) is also related to the brittleness and ductility of materials. A compound is considered brittle if the ν is <0.26 [29]. The higher value of Poisson's ratio is, the more ductile the material is. Thus, Tables 6–8 show that all REMO$_4$ materials are ductile. They are in good agreement with the results estimated by the B/G ratio. Besides, the value of Poisson's ratio suggests that the ductility is inversely proportional to the rare earth atom of REMO$_4$ materials, and RENbO$_4$ is more ductile than RETaO$_4$.

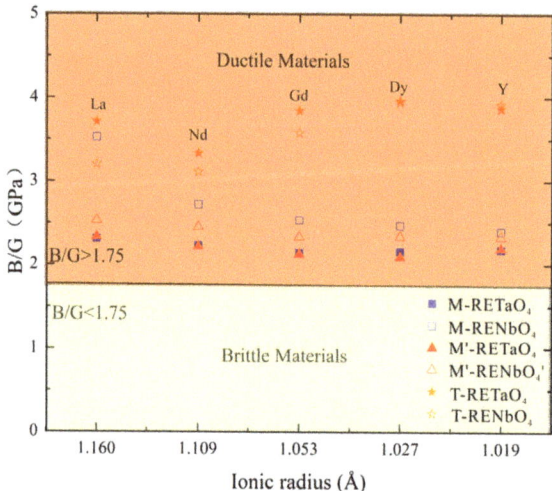

Figure 6. Calculated B/G of REMO$_4$ (M = Ta, Nb) along with the ionic radius of RE^{3+}.

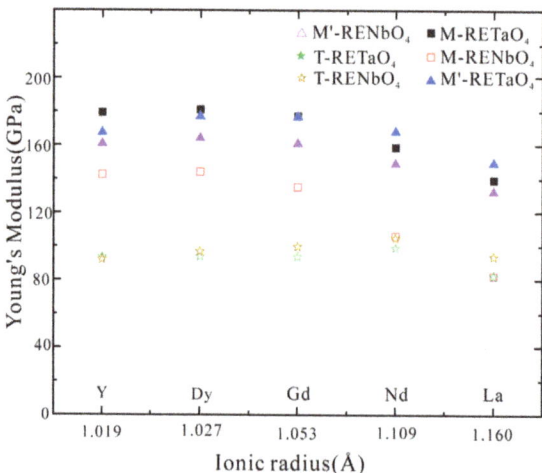

Figure 7. Calculated Young's modulus of M-, M'-, T-REMO$_4$ (M = Ta, Nb) along with the ionic radius of RE^{3+}.

Hardness is a very important mechanical property in applications. Hardness is defined as the resistance of a material to deformation and may be predicted using macroscopic and microscopic models. In this work, we use the semi-empirical equations of hardness proposed by Chen et al. [30] and Tian et al. [31] were used to study the hardness of the REMO$_4$ phases. The equations of these two models are defined as follow:

$$H_V = 2(k^2 G)^{0.585} - 3 \tag{10}$$

$$H_V = 0.92 k^{1.137} G^{0.708} \tag{11}$$

where k = G/B, G and B are the shear modulus and the bulk, respectively. The obtained hardness of REMO$_4$ phases are shown in Figure 8 and Tables 6–8 presents a comparison between the calculated and experimental results, which exhibits a good consistency. Besides, our calculations suggest that the Vickers hardness of the T phase decreases, and the single-phase gradually increase with the decrease of the atomic radius.

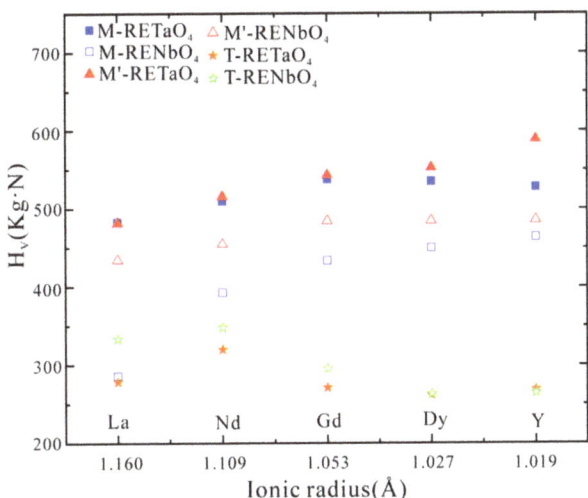

Figure 8. The calculated hardness of REMO$_4$ phases (M = Ta, Nb) along with the ionic radius of RE^{3+}.

The anisotropic mechanical properties of the compounds are very important in applications. Based on the G and B values from Reuss and Voigt, Ranganathan et al. [32] proposed a universal elastic anisotropy index A^U for crystal with any symmetry as shown below:

$$A^U = \frac{5G_{Voigt}}{G_{Reuss}} + \frac{B_{Voigt}}{B_{Reuss}} - 6 \geq 0 \tag{12}$$

A^U is equal to zero when the single crystals are locally isotropic. The extent of single-crystal anisotropy can be expressed by the departure from zero indicates. The highly mechanical anisotropic properties exhibit large discrepancies from zero. The calculated elastic anisotropy is shown in Figure 9. Most values of A^U are lower than 1. The larger the value of A^U is, the stronger the anisotropy of the phase. The M-, M′- and T-REMO$_4$ are anisotropic, and the elastic anisotropy of M and M′ phases is larger than the tetragonal (T) phase.

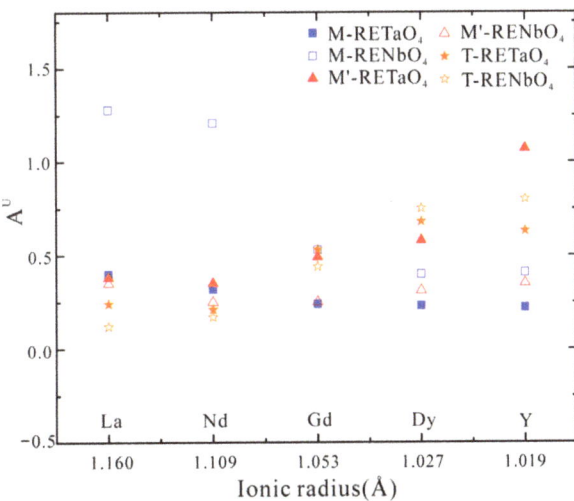

Figure 9. Calculated elastic anisotropy of REMO$_4$ phases (M = Ta, Nb) along with the ionic radius of RE^{3+}.

4. Conclusions

In the present work, phase stability and mechanical properties of REMO$_4$ (RE = La, Nd, Gd, Dy, Y; M = Ta, Nb) are investigated by first-principles calculations. Some conclusions can be found. For RETaO$_4$, the M' phase is more stable than the M phase at low temperature, and the T phase is only stable at high temperature. For RENbO$_4$, only the M phase is stable at low temperatures. This is consistent with the experimental results. Our calculated relative energies ($\Delta E_{M' \to T \text{ or } M \to T}$) of M'- or M-REMO$_4$ with respect to that of T-REMO$_4$ for different rare-earth atoms are inversely proportional to the ionic radius of RE^{3+}. This implies that the phase transformation temperature of M'→T or M→T is decreased with the increase of the rare-earth atoms, which is consistent with the experimental data. Moreover, our calculations exhibit that adding Nb into M'-RETaO$_4$ can induce phase transformation of M'→M, and the doping concentration is about 50%. Besides, the elastic coefficient is attained by means of the strain-energy method. Bulk modulus, Young's modulus, shear modulus, and Poisson's ratio of T-, M-, and M' phases are obtained according to Voigt-Reuss-Hill approximation, and anisotropic mechanical properties of the REMO$_4$ materials have been calculated. Finally, our calculated B/G exhibits that T-, M-, and M' phases are all ductile, and the hardness of REMO$_4$ phases are predicted based on semi-empirical equations, which is consistent with the experimental data.

Author Contributions: Investigation, writing—original draft preparation, W.X.; data curation, Y.Y.; funding acquisition, Z.P.; writing—review and editing, funding acquisition, F.Z. All authors have read and agreed to the published version of the manuscript.

Funding: This research was funded by the National Natural Science Foundation of China (No. 51801170 and 11802257), National Postdoctoral Program for Innovative Talents (No. BX20180265), Natural Science Foundation of Hunan Province (No. 2019JJ50570), China Postdoctoral Science Foundation (No. 2019M652786), Research initiation project of Xiangtan University (No. 18QDZ24) and Research foundation of education bureau of Hunan province, No. 21B0163.

Institutional Review Board Statement: Not applicable.

Informed Consent Statement: Not applicable.

Data Availability Statement: Not applicable.

Conflicts of Interest: The authors declare no conflict of interest.

References

1. Voloshyna, O.; Neicheva, S.V.; Starzhinskiy, N.G.; Zenya, I.M.; Gridin, S.S.; Baumer, V.N.; Sidletskiy, O.T. Luminescent and scintillation properties of orthotantalates with common formulae RETaO$_4$ (RE = Y, Sc, La, Lu and Gd). *Meter. Sci. Eng. B* **2013**, *178*, 1491–1496. [CrossRef]
2. Shian, S.; Sarin, P.; Gurak, M.; Baram, M.; Kriven, W.M.; Clarke, D.R. The tetragonal-monoclinic, ferroelastic transformation in yttrium tantalate and effect of zirconia alloying. *Acta Mater.* **2014**, *69*, 196–202. [CrossRef]
3. Feng, J.; Shian, S.; Xiao, B.; Clarke, D.R. First-principles calculations of the high-temperature phase transformation in yttrium tantalate. *Phys. Rev. B* **2014**, *90*, 094102. [CrossRef]
4. Flamant, Q.; Gurak, M.; Clarke, D.R. The effect of zirconia substitution on the high-temperature transformation of the monoclinic-prime phase in yttrium tantalate. *J. Eur. Cera. Soc.* **2018**, *38*, 3925–3931. [CrossRef]
5. Virkar, A.V. Role of Ferroelasticity in toughening of zirconia ceramics. *Key Eng. Mater.* **1998**, *153–154*, 183–210. [CrossRef]
6. Zhang, F.; Zhang, G.J.; Yang, L.; Zhou, Y.C.; Du, Y. Thermodynamic modeling of YO$_{1.5}$-TaO$_{2.5}$ system and the effects of elastic strain energy and diffusion on phase transformation of YTaO$_4$. *J. Eur. Ceram. Soc.* **2019**, *39*, 5036–5047. [CrossRef]
7. Brixner, L.H.; Chen, H.Y. On the structural and luminescent properties of the M'-LnTaO$_4$ rare-earth tantalates. *J. Electrochem. Soc.* **1983**, *130*, 2435–2443. [CrossRef]
8. Wang, J.; Yu, X.; Zhou, C.R.; Feng, J. Microstructure and thermal properties of RETaO$_4$ (RE = Nd, Eu, Gd, Dy, Er, Yb, Lu) as promising thermal barrier coating materials. *Scr. Mater.* **2017**, *126*, 24–28. [CrossRef]
9. Chen, L.; Wang, J.; Feng, J. Research progress of rare earth tantalate ceramics as thermal barrier coatings. *Sci. China Mater.* **2017**, *36*, 938–949.
10. Zhu, J.T.; Lou, Z.H.; Zhang, P.; Zhao, J.; Meng, X.Y.; Xu, J.; Gao, F. Preparation and thermal properties of rare earth tantalates (RETaO$_4$) high-entropy ceramics. *J. Inorg. Mater.* **2021**, *36*, 411–417. [CrossRef]

11. Kresse, G.; Furthmüller, J. Efficient iterative schemes for ab initio total-energy calculations using a plane-wave basis set. *Phys. Rev. B* **1996**, *54*, 11169–11186. [CrossRef] [PubMed]
12. Kresse, G.; Furthmüller, J. Efficiency of ab initio total energy calculations for metals and semiconductors using a planewave basis set. *Comp. Mater. Sci.* **1996**, *6*, 15–50. [CrossRef]
13. Blöchl, P.E. Projector augmented-wave method. *Phys. Rev. B* **1994**, *50*, 17953–17979. [CrossRef] [PubMed]
14. Perdew, J.P.; Burke, K.; Ernzerhof, M. Generalized gradient approximation made simple. *Phys. Rev. Lett.* **1996**, *77*, 3865–3868. [CrossRef]
15. Blöchl, P.E.; Jepsen, O.; Andersen, O.K. Mproved tetrahedron method for brillouin-zone integrations. *Phys. Rev. B* **1994**, *49*, 16223–16233. [CrossRef] [PubMed]
16. Landau, L.D.; Lifshitz, E.M. *Theory of Elasticity Butterworth*; Heinemann: Oxford, UK, 1999.
17. Voigt, W. *Lehrburch der Kristallphysik*; Teubner: Leipzig, Germany, 1928.
18. Reuss, A. Calculation of flow limits of mixed crystals on the basis of plasticity of single crystals. *Z. Angew. Math. Mech.* **1929**, *9*, 49–58. [CrossRef]
19. Hill, R. The elastic behaviour of a crystalline aggregate. *Proc. Phys. Soc. Sect. A* **1952**, *65*, 349–354. [CrossRef]
20. Sych, A.B.; Golub, M. Niobates and tantalates of tervalent elements. *Russ. Chem. Rev.* **1977**, *46*, 210. [CrossRef]
21. Hartenbach, I.; Lissner, F.; Nikelski, T.; Meier, S.F.; Müller-Bunz, H.; Schleid, T. Über oxotantalate der lanthanide des formeltyps MTaO$_4$ (M = La–Nd, Sm–Lu). *Z. Anorg. Allg. Chem.* **2005**, *631*, 2377–2382. [CrossRef]
22. Mather, S.A.; Davies, P.K. Nonequilirbium phase formation in oxides prepared at low temperature: Fergusonite-related phases. *J. Am. Ceram. Soc.* **1995**, *78*, 2737–2745. [CrossRef]
23. Zhang, F.; Wang, J.C.; Liu, S.H.; Du, Y. Effects of the volume changes and elastic-strain energies on the phase transition in the Li-Sn battery. *J. Power Sources* **2016**, *330*, 111–119. [CrossRef]
24. Stubičan, V.S. High-temperature transitions in rare-earth niobates and tantalates. *J. Am. Ceram. Soc.* **1964**, *47*, 55–58. [CrossRef]
25. Watt, J.P.; Peselnick, L. Clarification of the hashine shtrikman bounds on the effective elastic moduli of polycrystals with hexagonal, trigonal, and tetragonal symmetries. *J. Appl. Phys.* **1980**, *51*, 1525–1531. [CrossRef]
26. Söderlind, P.; Klepeis, J.E. First-principles elastic properties of α-Pu. *Phys. Rev. B* **2009**, *79*, 104110. [CrossRef]
27. Born, M. On the stability of crystals. *Proc. Camb. Philos. Soc.* **1940**, *36*, 160. [CrossRef]
28. Pugh, S.F. XCII. Relations between the elastic moduli and the plastic properties of polycrystalline pure metals. *Lond. Edinb. Dublin Philos. Mag. J. Sci.* **1954**, *45*, 823–843. [CrossRef]
29. Wei, N.; Zhang, X.L.; Zhang, C.G.; Hou, S.J.; Zeng, Z. First-principles investigations on the elastic and thermodynamic properties of cubic ZrO2 under high pressure. *Int. J. Mod. Phy. C* **2015**, *26*, 1550056. [CrossRef]
30. Chen, X.Q.; Niu, H.; Li, D.; Li, Y. Modeling hardness of polycrystalline materials and bulk metallic glasses. *Intermetallics* **2011**, *19*, 1275–1281. [CrossRef]
31. Tian, Y.; Xu, B.; Zhao, Z. Microscopic theory of hardness and design of novel superhard crystals. *J. Meta. Hard Mater.* **2012**, *33*, 93–106. [CrossRef]
32. Ranganathan, S.I.; Ostoja-Starzewski, M. Universal Elastic Anisotropy Index. M. *Phys. Rev. Lett.* **2008**, *101*, 055504. [CrossRef] [PubMed]

Article

Comparative Investigation on Corrosion Resistance of Stainless Steels Coated with Titanium Nitride, Nitrogen Titanium Carbide and Titanium-Diamond-like Carbon Films

Jia Lou [1], Zonglong Gao [1], Jie Zhang [2,3,*], Hao He [4,*] and Xinming Wang [1]

1. School of Materials Science and Engineering, Xiangtan University, Xiangtan 411105, China; lou3166@xtu.edu.cn (J.L.); gaozonglong1996@163.com (Z.G.); wangxm@xtu.edu.cn (X.W.)
2. School of Stomatology, Hunan University of Chinese Medicine, Changsha 410036, China
3. Changsha Stomatological Hospital, Changsha 410000, China
4. School of Microelectronics and Materials Engineering Research Centre for Materials Science and Engineering, Guangxi University of Science and Technology, Liuzhou 545006, China
* Correspondence: kqzj@hnucm.edu.cn (J.Z.); 100001865@gxust.edu.cn (H.H.)

Citation: Lou, J.; Gao, Z.; Zhang, J.; He, H.; Wang, X. Comparative Investigation on Corrosion Resistance of Stainless Steels Coated with Titanium Nitride, Nitrogen Titanium Carbide and Titanium-Diamond-like Carbon Films. *Coatings* 2021, *11*, 1543. https://doi.org/10.3390/coatings11121543

Academic Editor: Alina Vladescu

Received: 19 November 2021
Accepted: 13 December 2021
Published: 15 December 2021

Publisher's Note: MDPI stays neutral with regard to jurisdictional claims in published maps and institutional affiliations.

Copyright: © 2021 by the authors. Licensee MDPI, Basel, Switzerland. This article is an open access article distributed under the terms and conditions of the Creative Commons Attribution (CC BY) license (https://creativecommons.org/licenses/by/4.0/).

Abstract: In this study, the corrosion resistance of titanium nitride (TiN), nitrogen titanium carbide (TiCN) and titanium-diamond-like carbon (Ti-DLC) films deposited on 316L stainless steel (SS) were compared via differences in the surface and section-cross morphologies, open circuit potential tests, electrochemical impedance spectroscopy and potentiometric tests. The corrosion resistance of the TiCN and Ti-DLC films significantly improved because of the titanium carbide (TiC) crystals that obstruct the corrosive species penetrating the as-deposited film in the electrolyte atmosphere. TiN exhibited the lowest corrosion resistance because of its low thickness and high volume of defects. The Ti-DLC film showed the lowest corrosion current density (approximately 4.577 $\mu A/cm^2$) and thickness reduction (approximately 0.12 μm) in different electrolytes, particularly those with high Cl^- and H^+ concentrations, proving to be the most suitable corrosion protection material for 316L SS substrates.

Keywords: dental brace; anticorrosive coating; electrochemical test; titanium-diamond-like carbon film; titanium carbide cluster

1. Introduction

Dental braces are frequently used in the field of orthodontics because they arrange misplaced teeth in their normal positions, leading to improved oral health [1–3]. Conventionally used orthodontic appliances are mainly made of stainless steel (SS), which has a favorable combination of mechanical properties, corrosion resistance and cost effectiveness [4]. However, stainless steel is susceptible to wear and corrosion in an oral environment with changing pH values, high concentration of Cl^- and other ions, such as Na^+, K^+ and F^- [5–8]. Friction or collision causes stress corrosion, which accelerates brace failure and leads to the release of harmful nickel ions [9–11].

Anticorrosive coatings are often used to improve the corrosion resistance and wear resistance of dental materials to resolve these issues. TiN coatings demonstrate a high hardness, low coefficient of friction (COF), and high biocompatibility [12,13]. Liu et al. found that the corrosion current densities of TiN-coated specimens were approximately three orders of magnitude lower than those of uncoated specimens, indicating that TiN could protect substrates [14]. In contrast, Kao et al. revealed that TiN did not increase the anticorrosion ability of standard dental bracket because TiN-plated and non-TiN-plated brackets released detectable ions into the test solution. These included Ni^{2+}, Cr^{3+} and Fe^{3+}, suggesting differences in the porosity of the coatings [15]. TiCN coatings have attracted attention due to low internal stress, excellent tribological properties and non-toxicity [16–18]. Numerous studies have shown that doping C in TiN has a double-sided effect [19–21]. Wang et al. reported that when the C content of TiCN is 2.05%, the charge transfer resistance (R_{ct}) increases from 8.8×10^5 $\Omega \cdot cm^2$ in TiN to 6.49×10^6 $\Omega \cdot cm^2$ due to formation of α-CN_x, leading to a more stable electrochemical performance. When the C content increases to 2.46%, α-C is formed, deteriorating the corrosion resistance of TiCN in simulated body fluids [22]. With the increasing demand for comfortable and durable dental braces, diamond-like carbon (DLC) films are extensively being used in protective coatings for orthodontic materials. Ti-DLC films are formed through the co-deposition of DLC and biocompatible Ti. This process not only enhances the mechanical properties of the surface, such as hardness, toughness and tribological performance, but also improves its corrosion resistance, particularly stress corrosion resistance [23–26].

The tribological performances of 316L SS substrates coated with TiN, TiCN, and Ti-DLC films were compared in our previous studies. The Ti-DLC film showed the lowest COF because the surface contains sp^{2-} carbon (graphite-like carbon) that has a self-lubricating effect [27]. In addition, the Ti-DLC coatings demonstrated excellent corrosion resistance and electrochemical stability. Zhang et al. reported that Ti-DLC has a higher corrosion potential and lower corrosion current than those of the SS substrate, indicating its superior ability to protect the substrate [28]. Konkhunthot et al. also reported that the corrosion current was reduced by two orders of magnitude owing to formation of a Ti-DLC film on 304 SS [29]. More recently, Wongpanya et al. incorporated and interlayered Ti into Ti-DLC film, finding that the double-interlayer enhanced pitting corrosion resistance of the film [30]. Zhou et al. reported that excessive TiC acted as defect points in DLC films and weakened the corrosion resistances of the film. They observed a higher corrosion current density and a lower corrosion potential in Ti-DLC films with Ti contents of 5.31 at.% compared with 0.46 at.% [31].

The Ti-DLC film in our previous work had a modest hardness, the lowest roughness and COF; however, electrochemical corrosion tests were not performed. Studies comparing the corrosion mechanism of TiN, TiCN and Ti-DLC coatings under artificial saliva condition are rare, particularly with fluctuating pH value and Cl^- concentrations. In this study, 316L SS substrates coated with TiN, TiCN and Ti-DLC were prepared by multi-arc ion plating. Open circuit potential (OCP) tests, electrochemical impedance spectroscopy (EIS) observations, potentiometric polarization measurements, and corrosion morphology tests were conducted in artificial saliva and electrolyte with different pH and Cl^- concentrations, and the film showing the best anticorrosion performance was identified. This study provides an experimental and theoretical basis for the production, application and improvement of orthodontic braces in future.

2. Experiment

The 316L stainless steel plates (approximate size of $15 \times 15 \times 2$ mm^3) were sanded and polished using sandpaper (#200–#2000) prior to deposition. These polished surfaces were ultrasonically cleaned in ethanol for 10 min and deionized water for 5 min, to remove residual liquid. TiN, TiCN and Ti-DLC films were deposited on the surface of these plates via multi-arc ion plating. The details of the materials and parameters of coating processes are obtained from our previous studies [27].

The surface morphologies and thickness of the films after the electrochemical tests were observed using field-emission scanning electron microscopy (FEI Nova NanoSEM230, Chicago, IL, USA).

The electrochemical properties of the samples were characterized using CH660e electrochemical workstation (ChenHua, Shanghai, China). The traditional three-electrode cell system was used to examine the electrochemical behavior of the samples; a 2.25 cm^2 area of the sample was exposed as the working electrode; an Ag/AgCl electrode acted as the reference electrode, and a titanium slice was used as the counter electrode. The samples were immersed in an artificial saliva solution for 4000 s to evaluate the OCP at 37 °C. During EIS, the measurement frequency was swept from 10^{-2} to 10^5 Hz, and a sinusoidal disturbance voltage of 10 mV was applied. The EIS data were imported into

ZSimpWin software for further fitting and analysis of the corresponding equivalent circuit (EC). The scanning potential ranged from −1.5 V vs. Ag/AgCl to 1.5 V vs. Ag/AgCl at a scanning rate of 2 mV/s was used for obtaining the potentiometric polarization curves. The corrosion current density of the samples was estimated from the polarization curves using the Tafel extrapolation fitting method. Inhibition efficiencies (IE, η) were calculated by fitted corrosion current densities (I_{corr}) based on the following equation:

$$\eta(\%) = (1 - \frac{I_{corr}}{I_{corr}^0}) \times 100\%$$

where I_{corr} and I_{corr}^0 are current densities of treated and naked steel, respectively.

The EIS and potentiometric polarization measurements were performed in the artificial saliva solution, those contents were determined according to the standard of the corrosion resistance examination of SS and NiTi orthodontic wires [32], as illustrated in Table 1. The pH of the artificial saliva solution was approximately 6.8.

Table 1. Ingredients of the artificial saliva solution.

Ingredient	NaCl	KCl	$CaCl_2$	K_2PO_4	KSCN	CH_4N_2O
Content (mg/L)	400	400	795	690	300	1000

To evaluate the effects of pH values and Cl^- concentrations on the corrosion resistance, potentiometric polarization curve tests were performed in solutions with different NaCl, citric acid (H_3Cit) and potassium citrate (K_3Cit) concentrations, as given in Table 2. The method of operation is already described.

Table 2. Electrolyte concentrations in 1 L electrolyte solutions used for electrochemical measurements.

Concentration (g/L)	(0.9%, 6.6)	(0.9%, 3)	(3%, 6.6)	(3%, 3)
NaCl (mL)	9	9	30	30
0.1 M H_3Cit (mL)	5.6	74.4	5.6	74.4
0.1 M K_3Cit (mL)	74.4	5.6	74.4	5.6

For easy understanding, the notation (X%, Y) is used to represent the electrolyte solution with an NaCl concentration of X% and a pH value of Y.

Before conducting the electrochemical tests, the samples were immersed in artificial saliva for 24 h, to ensure that their surfaces were completely passivated. All the electrochemical tests were repeated at least thrice, and showed roughly the same values and trends.

3. Results and Discussion

3.1. Surface Morphologies and Composition Analysis

Our previous study [27] showed the as-prepared surfaces of 316L SS substrate was smooth, and the surface of TiN had some particles and pinholes. Moreover, TiCN and Ti-DLC had many nanocrystal clusters after physical vapor deposition (PVD). According to the XRD data, these clusters were the TiC which was generated by the reaction of Ti and C during deposition. Additionally, the thicknesses of the TiN, TiCN, and Ti-DLC films were 2.05, 4.10 and 4.48 µm, respectively. The Ti-DLC was found to exhibit the best adhesion and anti-wear properties.

Figure 1 demonstrates the surface morphologies of the 316L SS substrate and three films after the electrochemical tests in artificial saliva. The substrate surface is generally smooth, and only the diameter of the holes has increased, owing to the low-degree erosion of SS in a relatively mild saliva environment. Furthermore, the surfaces of all three films have nanocrystal clusters and pinholes. This because the TiC particles exist in the TiCN and Ti-DLC films, functioning as seeds and gradually growing into cluster defects during deposition. In TiN samples, particles and pinholes are common growth defects resulting from deposition, which are always generated in the bulges and cavities on the substrate [33]. However, the number of particles of the TiN surface are far more than those of TiCN and Ti-DLC films after polarization tests in artificial saliva. This phenomenon is mainly because of the superior electrochemical stability of TiN clusters compared with those of TiC on Ti-DLC films; thus, TiC preferably dissolves in artificial saliva [34].

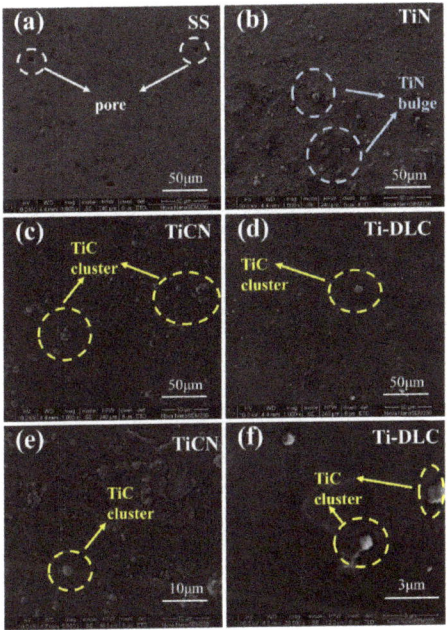

Figure 1. Surface morphologies of four samples after polarization tests in artificial saliva. (**a**) 316L SS, 1000×; (**b**) TiN, 1000×; (**c**) TiCN, 1000×; (**d**) Ti-DLC, 1000×; (**e**) TiCN, 5000×; (**f**) Ti-DLC, 20,000×.

3.2. OCP Test

Figure 2 displays the OCP variation curves of the four samples in artificial saliva at 37 °C within 4000 s. According to the standard test method for measuring potentiodynamic polarization resistance, the basic stabilization value of OCP can be used as valid data after 55 min. The 316L SS substrate shows the most negative OCP (−0.156 V). In contrast, the three films have higher potentials, indicating that all the films can significantly enhance the corrosion resistance stability of 316L SS substrate. Among them, TiCN possess the most positive electrode potential (0.092 V) and highest stability. Moreover, the Ti-DLC film shows a slight negative deviation in the OCP value, indicating the inferior chemical stability of the coating. This phenomenon results in a dense surface structure and relatively high thickness of the TiCN film, yielding excellent electrochemical stability. Although Ti-DLC has the highest thickness, it contains many clusters that could have dissolved in the artificial saliva, leading to a negative deviation in the OCP value. The TiN film has a lower potential and larger fluctuation than other films. Its inferior chemical stability is because of its low thickness and several defects.

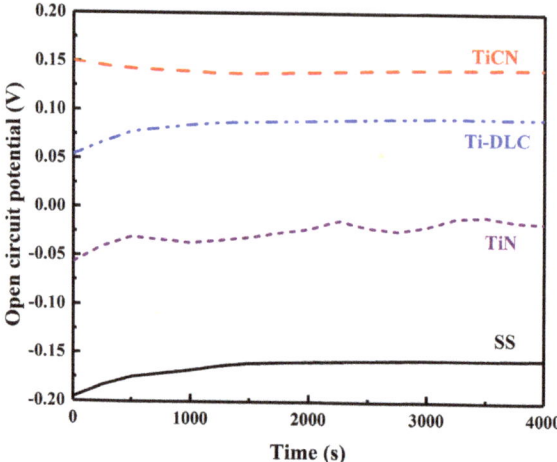

Figure 2. OCP of four samples in artificial saliva at 37 °C.

3.3. EIS Results

To evaluate the corrosion protection properties of the films, EIS measurements were carried out in artificial saliva at 37 °C. The Nyquist (including the locally amplified inner graphic) and Bode plots of the samples are shown in Figure 3a,b, respectively. The diameter of the capacitive loop in the Nyquist plots corresponds to the corrosion resistance. Figure 3a demonstrates that the corrosion resistances of the SS substrate, TiCN, Ti-DLC and TiN film decrease successively. The same order is found for the "log Z" values in Figure 3b. According to Zhang's research, the impedance modulus at the lowest frequency (Z_f = 0.01 Hz) is another parameter in the Bode plot for evaluating the corrosion protection capability of the films; a greater impedance suggests a better corrosion resistance [35].

In addition, the "phase angle" of the Bode plot and the locally amplified graphic of 10–10k Hz are illustrated in Figure 3c,d. As shown in Figure 3c, the order of the sample corrosion resistance (Figure 3c) was consistent with the Nyquist plots, owing to the improvement of the passive film, that increased the phase angle [36]. Figure 3c and d show the 316L SS has only one plateau region and all the films have two; this indicates that the EC of the 316L SS has one time constant, but the three films have two.

Considering a passive film formed on the 316L SS surface and electrochemical reactions took place at the metal/film interface, the best fitting EC simulated from the EIS results using the ZSimpWin software, is illustrated in Figure 4. Additionally, the parameters are listed in Table 3; the chi-squared (X^2) value in the order of 10^{-3} to 10^{-4} indicating the good fit has been obtained with the suggested EC models [37]. This method has been used to fit titanium alloys and metal-DLC films in many studies [22,38,39]. In the circuits (a), the R_s, R_1 and R_2 represented the solution resistance, porous layer resistance, and the dense layer resistance, respectively. In the circuits (b), the R_s, R_f and R_{ct} represented the solution resistance, film resistance and the charge transfer resistance, respectively. The C_1 and C_2 were the surface layer capacitance and inner layer capacitance; additionally, C_{dl} represented the double-layer capacitance. In Figure 4a, R_2 is 25,320 $\Omega \cdot cm^2$ indicates that the oxide layer produced via the self-passivation of the SS substrate can effectively improve its corrosion resistance. In Figure 4b, the maximum R_f (637.8 $\Omega \cdot cm^2$) was observed in TiCN, indicating its excellent corrosion resistance. R_f of Ti-DLC and TiN are 329.1 $\Omega \cdot cm^2$ and 154.9 $\Omega \cdot cm^2$, respectively. It worth noting that R_{ct} of TiCN is similar to that of Ti-DLC, implying that the charge transfer at both these substrate interfaces is effectively prevented, thus contributing to their high corrosion resistance. This enhancement in corrosion resistance may be related to the dense surface structure and relatively high thickness of TiCN. Furthermore, Ti-DLC has the highest thickness, but contains many clusters that dissolved into the artificial saliva;

thus, its R_f value is smaller than TiCN. Moreover, the main reason for the poor corrosion resistance of the TiN film is its low thickness and high volume of defects.

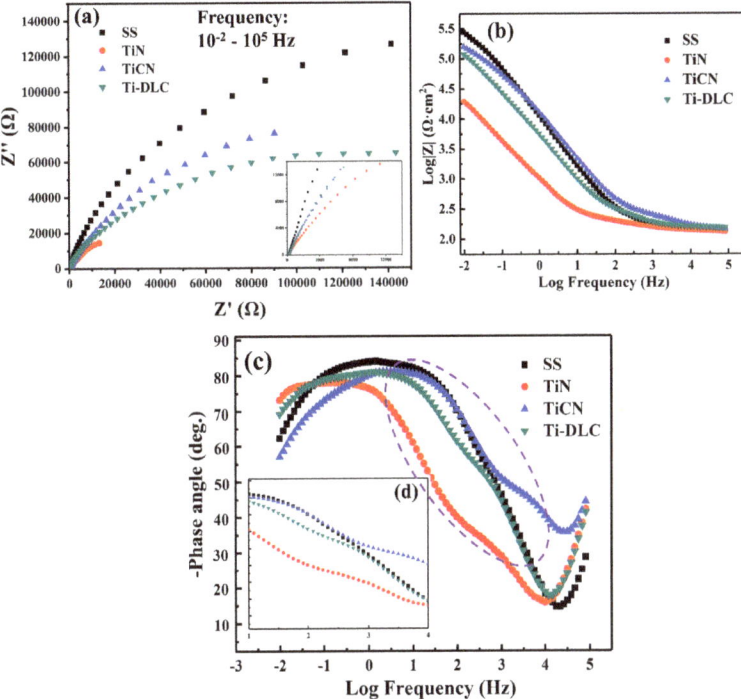

Figure 3. Nyquist and Bode plots of the four samples in artificial saliva. (**a**) Nyquist plots of the four samples; (**b**) Bode plots of the four samples; (**c**) Phase angle of bode plots; insertion (**d**) locally amplified graphic of phase angle.

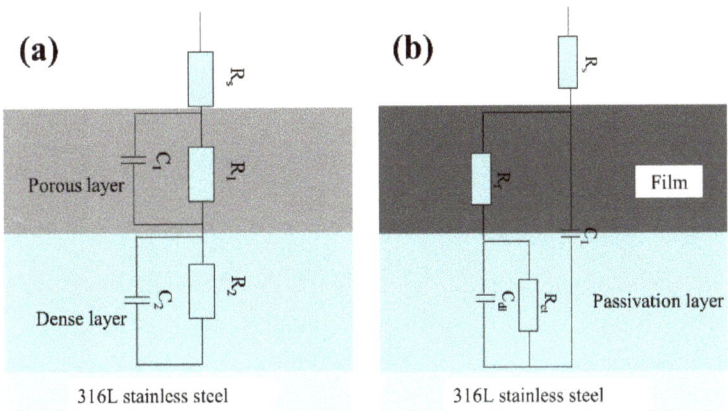

Figure 4. ECs of bare SS (**a**) and film-coated SS (**b**) used for fitting the impedance data.

Table 3. Thicknesses and EC results of 316L SS substrate, TiN, TiCN and, Ti-DLC films.

Sample	R_s ($\Omega \cdot cm^2$)	R_1/R_f ($\Omega \cdot cm^2$)	R_2/R_{ct} ($\Omega \cdot cm^2$)	C_1 (F/cm^2)	C_2/C_{dl} (F/cm^2)	Chi-Squared (X^2)
SS	17.85	218.3	25,320	1.297×10^{-4}	1.775×10^{-4}	2.3×10^{-3}
TiN	17.82	154.9	2230	9.884×10^{-4}	3.251×10^{-3}	7.2×10^{-4}
TiCN	21.16	637.8	11,760	1.404×10^{-4}	3.959×10^{-3}	4.5×10^{-3}
Ti-DLC	20.22	329.1	10,250	1.574×10^{-4}	3.240×10^{-4}	5.3×10^{-4}

3.4. Potentiometric Test

The Tafel polarization curves of the four samples immersed in artificial saliva are demonstrated in Figure 5a. A pronounced anodic peak emerged within the potential range from 0.592 to 0.618 V vs. Ag/AgCl and 0.185 to 0.223 V vs. Ag/AgCl, for 316L SS and TiN, respectively. The anodic peak was attributed to the film erosion in artificial saliva because the corrosive mediums can penetrate defects and microcracks of the film [40]. For Ti-DLC, it possessed a smooth anodic curve and a negative passive region, suggesting the less corrosion [26]. Normally, a low corrosion current density (I_{corr}) implies high corrosion-protection performance distinctly [37]. The I_{corr} was calculated using Tafel extrapolation and is summarized in Figure 5b. The Ti-DLC film possesses the lowest I_{corr} in artificial saliva, indicating the best corrosion resistance. This result is consistent with the polarization curve in Figure 5a. Its I_{corr} is considerably lower than that of the TiCN film, due to its ultrahigh thickness and that its surface clusters have already dissolved already. Wang et al. also reported an increase in electrochemical stability after the surface clusters dissolved [22]. They found that the TiCN coating with many clusters on the surface showed poor stability during the EIS test. With the dissolution of its surface clusters, a smoother surface was formed, leading to better corrosion resistance in subsequent potentiometric tests. The I_{corr} of Ti-DLC is higher than other similar DLC coatings, which may be due to the larger surface roughness (10.4 nm) than the others (approximately 4.2 nm) in this study. Other studies have shown large surface roughness weakens the corrosion resistance [28,31]. In addition to film thickness and roughness, the corrosion resistance may be affected by the degree of crystallization and grain size. These parameters can be controlled by the substrate temperature, deposition speed, external electric field, gas flow rate, and pressure. This could be the focus of a future study. The polarization curve for TiN shows a passive region in a narrower potential range and the highest I_{corr}, demonstrating the formation of an inferior stable passive layer compared with those of the other films.

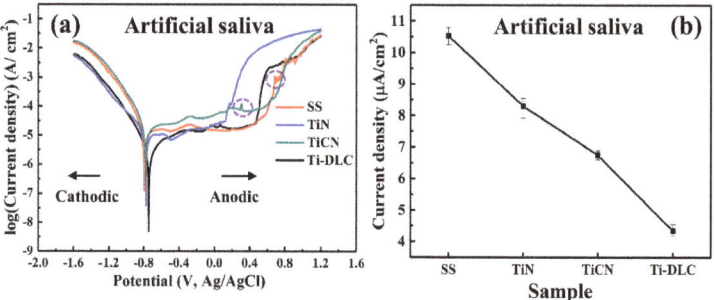

Figure 5. Polarization plots and corrosion current densities of the four samples in artificial saliva. (a) Polarization plots of the four samples; (b) corrosion current densities of the four samples.

The electrochemical behaviors of the samples in different electrolyte systems were analyzed to evaluate the effects of the oral microenvironment on corrosion resistance, and the data are presented in Figure 6. The performance of 316L SS is different from that discussed in Section 3.3. The polarization curves of the 316L SS shift toward a more positive region with increases in the Cl^- or H^+ concentrations. In addition, its I_{corr} increased significantly, suggesting that the corrosion resistance of 316L SS was significantly affected by the changes in the electrolyte systems. This is because 316L SS contains corrosion-resistant elements, namely Cr and Ni, capable of forming a dense oxide film on the surface of the material [41]; thus, the best corrosion resistance was obtained from the EIS results. However, at high concentrations of Cl^-, the Cl^- were preferentially adsorbed onto 316L SS surface because of their affinity to metals. The locally adsorbed Cl^- ions led to enhanced dissolution of the passive layer at these sites, thinning of the passive layer and eventually active dissolution. Further, the chemical activity of the H^+ cations significantly accelerated this process [42]. Figure 5e shows, the TiN films are insensitive to changes in pH. However, as the concentration of Cl^- increased from 0.9% to 3%, I_{corr} increased from 10.588 to 37.51 µA/cm². According to Liu, during the dissolution of TiN, the H^+ ions are neutralized by the N ions released from the TiN, transforming into NH_4^+ [43]. However, in a Cl^--rich atmosphere, the TiN could not prevent Cl^- from penetrating the film completely because of its low thickness and imperfections. Once the corrosive ions penetrate the coating through these small channels, a galvanic effect is formed at the interface between the coating and substrate, which accelerates the corrosion rate because of the difference in composition. Thus, I_{corr} of TiN is considerably higher than that of the 316L SS substrate in the (3%, 3) electrolyte [44]. The protection mechanism of the TiN film is illustrated in Figure 7a.

The TiCN and Ti-DLC films exhibited excellent electrochemical stability in different electrolytes. The main reason for this phenomenon is that the TiC crystals can obstruct the path of the corrosive species penetrating the as-deposited film. The corrosive medium hardly contacted the 316L SS substrate. Hence, the corrosion resistance of the bracket is improved [44,45]. However, note that I_{corr} of TiCN in the (3%, 6.6) electrolyte is significantly higher than that in other concentrations. A possible reason is that the thickness of the TiCN film is lower than that of the Ti-DLC film; This means several Cl^- and H^+ can easily penetrate the film/sublayer interface to react with the substrate. For Ti-DLC, the corrosion current increases slightly at concentrations of high Cl^- and H^+ ions, owing to its ultrahigh thickness and self-passivation effect on the electrolyte. Further corrosion can be prevented, thereby protecting the 316L SS substrate. The corrosion-resistant mechanism of the Ti-DLC film is shown in Figure 7b. All parameters of the four samples calculated using Tafel extrapolation are summarized in Table 4.

Table 4. All parameters of four sample calculated using Tafel extrapolation.

Electrolyte	Sample	E_{corr} (V)	I_{corr} (A)	βa (dec)	βa (dec)	η (%)
artificial	SS	−0.7936	1.05×10^{-5}	690	−134	-
	TiN	−0.7508	0.83×10^{-5}	106	−167	20.95
	TiCN	−0.7342	0.68×10^{-5}	225	−108	35.24
	Ti-DLC	−0.7292	0.44×10^{-5}	750	−225	58.09
0.9%, 6.6	SS	−0.7806	0.91×10^{-5}	125	−171	-
	TiN	−0.643	0.71×10^{-5}	225	667	21.98
	TiCN	−0.8038	0.55×10^{-5}	208	−113	39.56
	Ti-DLC	−0.425	0.28×10^{-5}	667	−417	69.23
0.9%, 3	SS	−0.6544	2.61×10^{-5}	125	−225	-
	TiN	−0.4787	1.06×10^{-5}	75	−105	59.39
	TiCN	−0.4528	0.98×10^{-5}	334	−149	62.45
	Ti-DLC	−0.4088	0.46×10^{-5}	183	−417	82.24
3%, 3	SS	−0.7711	2.92×10^{-5}	746	291	-
	TiN	−0.727	3.41×10^{-5}	91	−646	-
	TiCN	−0.5396	2.06×10^{-5}	159	−158	29.45
	Ti-DLC	−0.4571	0.61×10^{-5}	747	−118	79.11
3%, 6.6	SS	−0.86	1.96×10^{-5}	167	−91	-
	TiN	−0.8815	3.76×10^{-5}	375	201	-
	TiCN	−0.5373	1.54×10^{-5}	146	−267	21.43
	Ti-DLC	−0.4118	0.46×10^{-5}	167	−133	76.53

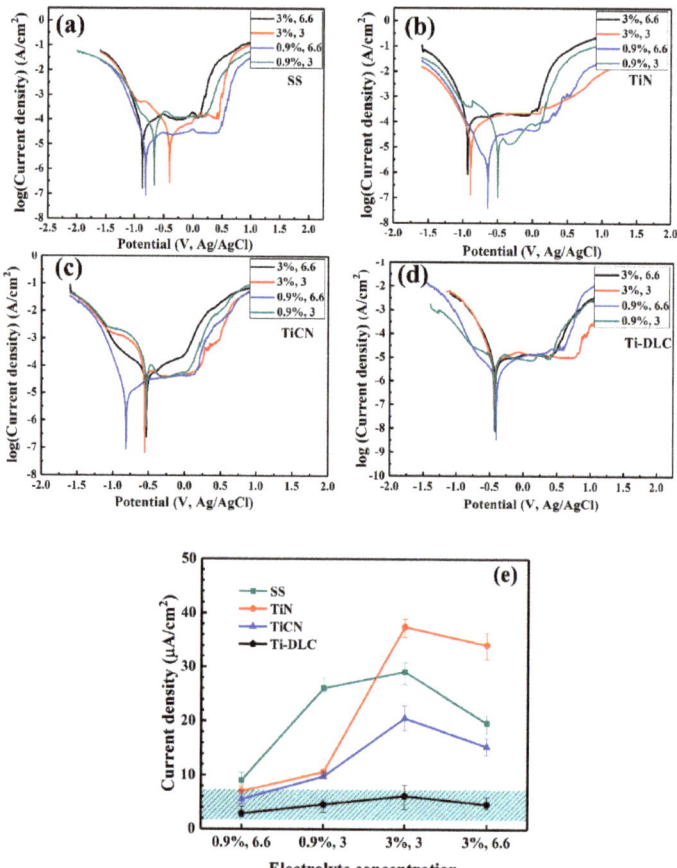

Figure 6. Polarization plots and corrosion current densities of the four samples in different electrolytes. (**a**) Polarization plots of SS; (**b**) Polarization plots of TiN; (**c**) Polarization plots of TiCN; (**d**) Polarization plots of Ti-DLC; (**e**) corrosion current densities of the four samples.

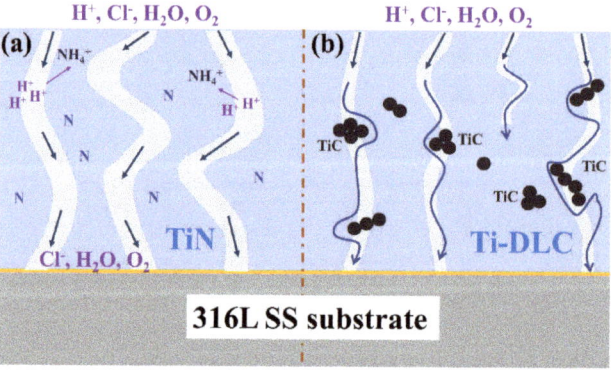

Figure 7. Protection mechanism of TiN and TiCN films in high H^+, Cl^- concentrate solution. (**a**) TiN; (**b**) Ti-DLC.

3.5. Surface Morphologies and Thickness of Sample after Electrochemical Tests

The above findings are consistent with the surface morphologies of the 316L SS substrate and three films after the electrochemical tests, as shown in Figure 8. The pore size of the 316L SS substrate surface significantly increases after electrochemical tests, particularly in the electrolyte system with high Cl^- and H^+ concentrations. Additionally, TiN and TiCN show fewer holes at low Cl^- concentrations. Simultaneously at high Cl^- and H^+ concentrations, noticeable pitting occurred on the TiN surface, and the TiCN surface became uneven. The surface of Ti-DLC was relatively smooth and complete, indicating that the Ti-DLC film could effectively reduce the corrosion tendency of 316L SS under various conditions.

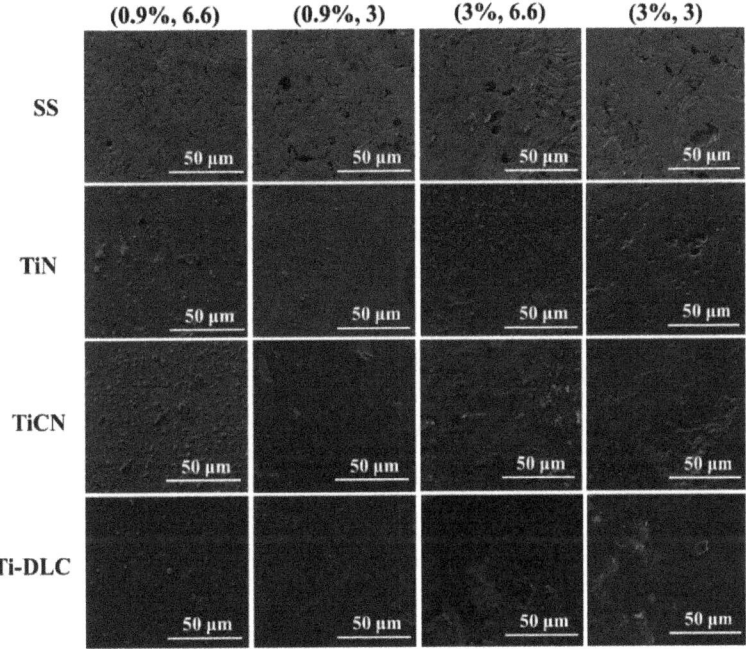

Figure 8. Surface morphologies of the four samples after electrochemical tests in different electrolytes.

Figure 9 demonstrates the cross-section morphologies of the three films after the electrochemical tests. The thicknesses of all films decrease by different amounts after the electrochemical tests. The thicknesses of TiN, TiCN and Ti-DLC reduce from 2.05, 4.10 and 4.48 µm to 1.69, 3.92 and 4.36 µm, respectively. In case of TiCN and Ti-DLC films, their thicknesses reduced by 0.18 and 0.12 µm, indicating that the corrosion level of Ti-DLC is lower. The thickness of the TiN film reduced by 0.36 µm, proving that it has the worst corrosion resistance.

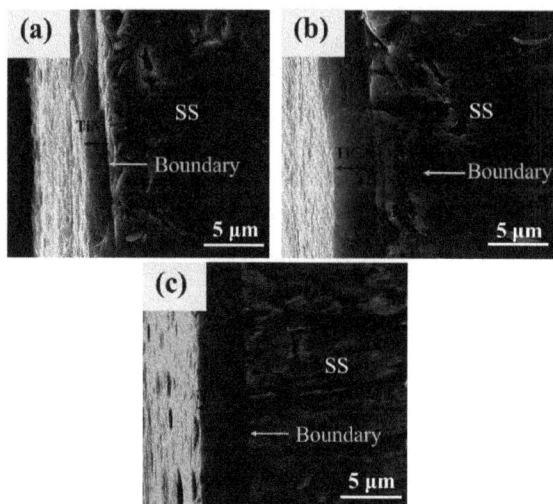

Figure 9. Cross-section morphologies of the three films after electrochemical tests. (**a**) TiN; (**b**) TiCN; (**c**) Ti-DLC.

4. Conclusions

In summary, we compared the corrosion mechanisms of TiN, TiCN and Ti-DLC coatings on 316L SS substrates to identify the best corrosion protection coating for dental braces. The main findings of the study are as follows:

(1) In the initial stage of an electrochemical tests, the stability of TiCN and Ti-DLC is inferior to 316L SS because the clusters on their surface dissolve preferentially in the artificial saliva.

(2) After the EIS tests, the clusters on TiCN and Ti-DLC film dissolved and a smooth surface was obtained. The potentiometric test results showed that the Ti-DLC film exhibited the most stable electrochemical properties, regardless of the presence of artificial saliva or electrolytes with different Cl^- and H^+ concentrations. The excellent corrosion resistance of the Ti-DLC coating resulted from the TiC crystals obstructing the path of the corrosive species penetrating the as-deposited film.

(3) After the electrochemical tests, Ti-DLC had the smallest thickness reduction, implying that it had good chemical stability and a self-passivation effect in the electrolyte.

Author Contributions: Conceptualization, J.L. and J.Z.; methodology, J.Z.; software, Z.G.; validation, J.L., Z.G. and H.H.; formal analysis, J.Z.; investigation, J.Z.; resources, X.W.; data curation, Z.G.; writing—original draft preparation, Z.G.; writing—review and editing, J.L. and Z.G.; visualization, J.Z.; supervision, H.H.; project administration, J.Z.; funding acquisition, J.L., H.H. and X.W. All authors have read and agreed to the published version of the manuscript.

Funding: This research received no external funding by National Nature Science Foundation of China (52164042 and 51804271); The Natural Science Foundation of Hunan Province, China (2019JJ80055); Guangxi Natural Science Foundation (2018GXNSFAA281237); And Scientific Research Foundation of Hunan Provincial Education Department (19B550).

Institutional Review Board Statement: Not applicable.

Informed Consent Statement: Not applicable.

Data Availability Statement: Not applicable.

Conflicts of Interest: The authors declare that they have no conflict of interest.

References

1. Nattrass, C.; Sandy, J. Adult orthodontics—A review. *Br. J. Orthod.* **1995**, *22*, 331–337. [CrossRef]
2. Hu, H.; Li, C.; Li, F.; Chen, J.; Sun, J.; Zou, S.; Sandham, A.; Xu, Q.; Riley, P.; Ye, Q. Enamel etching for bonding fixed orthodontic braces. *Cochrane Database Syst. Rev.* **2013**, *1*, 1–10. [CrossRef]
3. Littlewood, S.J.; Millett, D.T.; Doubleday, B.; Bearn, D.R.; Worthington, H.V. Retention procedures for stabilising tooth position after treatment with orthodontic braces. *Cochrane Database Syst. Rev.* **2016**, *1*, 1–12. [CrossRef]
4. Rezaee, S.; Arman, A.; Jurečka, S.; Korpi, A.G.; Mwema, F.; Luna, C.; Sobola, D.; Kulesza, S.; Shakoury, R.; Bramowicz, M. Effect of annealing on the micromorphology and corrosion properties of Ti/SS thin films. *Superlattices Microstruct.* **2020**, *146*, 106681. [CrossRef]
5. Khelfaoui, Y.; Kerkar, M.; Bali, A.; Dalard, F. Electrochemical characterisation of a PVD film of titanium on AISI 316L stainless steel. *Surf. Coat. Technol.* **2006**, *200*, 4523–4529. [CrossRef]
6. Mirjalili, M.; Momeni, M.; Ebrahimi, N.; Moayed, M.H. Comparative study on corrosion behaviour of Nitinol and stainless steel orthodontic wires in simulated saliva solution in presence of fluoride ions. *Mater. Sci. Eng. C* **2013**, *33*, 2084–2093. [CrossRef]
7. Vaughan, J.L.; Duncanson, M.G., Jr.; Nanda, R.S.; Currier, G.F. Relative kinetic frictional forces between sintered stainless steel brackets and orthodontic wires. *Am. J. Orthod. Dentofac. Orthop.* **1995**, *107*, 20–27. [CrossRef]
8. Zhang, J.; Ju, P.; Wang, C.; Dun, Y.; Zhao, X.; Zuo, Y.; Tang, Y. Corrosion Behaviour of 316L Stainless Steel in Hot Dilute Sulphuric Acid Solution with Sulphate and NaCl. *Prot. Met. Phys. Chem.* **2019**, *55*, 148–156. [CrossRef]
9. Paetyangkul, A.; Türk, T.; Elekdağ-Türk, S.; Jones, A.S.; Petocz, P.; Cheng, L.L.; Darendeliler, M.A. Physical properties of root cementum: Part 16. Comparisons of root resorption and resorption craters after the application of light and heavy continuous and controlled orthodontic forces for 4, 8, and 12 weeks. *Am. J. Orthod. Dentofac. Orthop.* **2011**, *139*, e279–e284. [CrossRef]
10. Alfonso, M.; Espinar, E.; Llamas, J.M.; Rupérez, E.; Manero, J.; Barrera, J.; Solano, E.; Gil, F. Friction coefficients and wear rates of different orthodontic archwires in artificial saliva. *J. Mater. Sci. Mater. Med.* **2013**, *24*, 1327–1332. [CrossRef] [PubMed]
11. Jensen, C.S.; Lisby, S.; Baadsgaard, O.; Byrialsen, K.; Menné, T. Release of nickel ions from stainless steel alloys used in dental braces and their patch test reactivity in nickel-sensitive individuals. *Contact Dermat.* **2003**, *48*, 300–304. [CrossRef]
12. Jindal, P.; Santhanam, A.; Schleinkofer, U.; Shuster, A. Performance of PVD TiN, TiCN, and TiAlN coated cemented carbide tools in turning. *Int. J. Refract. Met. Hard Mater.* **1999**, *17*, 163–170. [CrossRef]
13. Datta, S.; Das, M.; Balla, V.K.; Bodhak, S.; Murugesan, V. Mechanical, wear, corrosion and biological properties of arc deposited titanium nitride coatings. *Surf. Coat. Technol.* **2018**, *344*, 214–222. [CrossRef]
14. Liu, C.; Chu, P.K.; Lin, G.; Qi, M. Anti-corrosion characteristics of nitride-coated AISI 316L stainless steel coronary stents. *Surf. Coat. Technol.* **2006**, *201*, 2802–2806. [CrossRef]
15. Kao, C.T.; Ding, S.J.; Chen, Y.C.; Huang, T.H. The anticorrosion ability of titanium nitride (TiN) plating on an orthodontic metal bracket and its biocompatibility. *J. Biomed. Mater. Res.* **2002**, *63*, 786–792. [CrossRef] [PubMed]
16. Takadoum, J.; Houmid-Bennani, H.; Mairey, D.; Zsiga, Z. Adhesion and wear resistance of thin hard coatings. *Eur. Ceram. Soc.* **1997**, *17*, 1929–1932. [CrossRef]
17. Ren, X.; Zhao, R.; Wang, W.; Song, X.; Zhang, Y.; Zhang, C. Corrosion resistance of TiCN films prepared with combining multi-arc ion plating and magnetron sputtering technique. *Rare Met. Mater. Eng.* **2018**, *47*, 2028–2036. [CrossRef]
18. Antunes, R.; Rodas, A.; Lima, N.; Higa, O.; Costa, I. Study of the corrosion resistance and in vitro biocompatibility of PVD TiCN-coated AISI 316 L austenitic stainless steel for orthopedic applications. *Surf. Coat. Technol.* **2010**, *205*, 2074–2081. [CrossRef]
19. Cheng, Y.; Browne, T.; Heckerman, B.; Meletis, E. Influence of the C content on the mechanical and tribological properties of the TiCN coatings deposited by LAFAD technique. *Surf. Coat. Technol.* **2011**, *205*, 4024–4029. [CrossRef]
20. Huang, S.; Ng, M.; Samandi, M.; Brandt, M. Tribological behaviour and microstructure of TiCxN (1− x) coatings deposited by filtered arc. *Wear* **2002**, *252*, 566–579. [CrossRef]
21. Senna, L.; Achete, C.; Hirsch, T.; Freire, F., Jr. Structural, chemical, mechanical and corrosion resistance characterization of TiCN coatings prepared by magnetron sputtering. *Surf. Coat. Technol.* **1997**, *94*, 390–397. [CrossRef]
22. Wang, Q.; Zhou, F.; Zhou, Z.; Li, L.K.-Y.; Yan, J. Electrochemical performance of TiCN coatings with low carbon concentration in simulated body fluid. *Surf. Coat. Technol.* **2014**, *253*, 199–204. [CrossRef]
23. Jo, Y.J.; Zhang, T.F.; Son, M.J.; Kim, K.H. Synthesis and electrochemical properties of Ti-doped DLC films by a hybrid PVD/PECVD process. *Appl. Surf. Sci.* **2018**, *433*, 1184–1191. [CrossRef]
24. Zhao, F.; Li, H.; Ji, L.; Wang, Y.; Zhou, H.; Chen, J. Ti-DLC films with superior friction performance. *Diam. Relat. Mater.* **2010**, *19*, 342–349. [CrossRef]
25. Qiang, L.; Zhang, B.; Zhou, Y.; Zhang, J. Improving the internal stress and wear resistance of DLC film by low content Ti doping. *Solid State Sci.* **2013**, *20*, 17–22. [CrossRef]
26. Xu, X.; Zhou, Y.; Liu, L.; Guo, P.; Li, X.; Lee, K.-R.; Cui, P.; Wang, A. Corrosion behavior of diamond-like carbon film induced by Al/Ti co-doping. *Appl. Surf. Sci.* **2020**, *509*, 144877. [CrossRef]
27. Zhang, J.; Lou, J.; He, H.; Xie, Y. Comparative investigation on the tribological performances of TiN, TiCN, and Ti-DLC film-coated stainless steel. *JOM* **2019**, *71*, 4872–4879. [CrossRef]
28. Zhang, S.; Yan, M.; Yang, Y.; Zhang, Y.; Yan, F.; Li, H. Excellent mechanical, tribological and anti-corrosive performance of novel Ti-DLC nanocomposite thin films prepared via magnetron sputtering method. *Carbon* **2019**, *151*, 136–147. [CrossRef]

29. Konkhunthot, N.; Photongkam, P.; Wongpanya, P. Improvement of thermal stability, adhesion strength and corrosion performance of diamond-like carbon films with titanium doping. *Appl. Surf. Sci.* **2019**, *469*, 471–486. [CrossRef]
30. Wongpanya, P.; Pintitraratibodee, N.; Thumanu, K.; Euaruksakul, C. Improvement of corrosion resistance and biocompatibility of 316L stainless steel for joint replacement application by Ti-doped and Ti-interlayered DLC films. *Surf. Coat. Technol.* **2021**, *425*, 127734. [CrossRef]
31. Zhou, Y.; Li, L.; Hu, T.; Wang, Q.; Shao, W.; Rao, L.; Xing, X.; Yang, Q. Role of TiC nanocrystalline and interface of TiC and amorphous carbon on corrosion mechanism of titanium doped diamond-like carbon films: Exploration by experimental and first principle calculation. *Appl. Surf. Sci.* **2021**, *542*, 148740. [CrossRef]
32. Pytko-Polonczyk, J.; Jakubik, A.; Przeklasa-Bierowiec, A.; Muszynska, B. Artificial saliva and its use in biological experiments. *J. Physiol. Pharmacol.* **2017**, *68*, 807–813.
33. Panjan, P.; Drnovšek, A.; Gselman, P.; Čekada, M.; Panjan, M. Review of growth defects in thin films prepared by PVD techniques. *Coatings* **2020**, *10*, 447. [CrossRef]
34. Cheng, Y.; Zheng, Y. Characterization of TiN, TiC and TiCN coatings on Ti–50.6 at.% Ni alloy deposited by PIII and deposition technique. *Surf. Coat. Technol.* **2007**, *201*, 4909–4912. [CrossRef]
35. Zhang, S.; Zhao, H.; Shu, F.; Wang, G.; Liu, B.; Xu, B. Study on the corrosion behavior of steel Q315NS heat-affected zone in a HCl solution using electrochemical noise. *RSC Adv.* **2018**, *8*, 454–463. [CrossRef]
36. Ye, C.; Hu, R.; Dong, S.; Zhang, X.; Hou, R.; Du, R.; Lin, C.; Pan, J. EIS analysis on chloride-induced corrosion behavior of reinforcement steel in simulated carbonated concrete pore solutions. *J. Electroanal. Chem.* **2013**, *688*, 275–281. [CrossRef]
37. Alves, M.M.; Prošek, T.; Santos, C.F.; Montemor, M.F. Evolution of the in vitro degradation of Zn–Mg alloys under simulated physiological conditions. *RSC Adv.* **2017**, *7*, 28224–28233. [CrossRef]
38. Bayón, R.; Igartua, A.; González, J.; De Gopegui, U.R. Influence of the carbon content on the corrosion and tribocorrosion performance of Ti-DLC coatings for biomedical alloys. *Tribol. Int.* **2015**, *88*, 115–125. [CrossRef]
39. Carnot, A.; Frateur, I.; Zanna, S.; Tribollet, B.; Dubois-Brugger, I.; Marcus, P. Corrosion mechanisms of steel concrete moulds in contact with a demoulding agent studied by EIS and XPS. *Corros. Sci.* **2003**, *45*, 2513–2524. [CrossRef]
40. Sun, W.; Wang, L.; Wu, T.; Wang, M.; Yang, Z.; Pan, Y.; Liu, G. Inhibiting the corrosion-promotion activity of graphene. *Chem. Mater.* **2015**, *27*, 2367–2373. [CrossRef]
41. Shih, C.; Shih, C.; Su, Y.; Su, L.H.; Chang, M.; Lin, S. Effect of surface oxide properties on corrosion resistance of 316L stainless steel for biomedical applications. *Corros. Sci.* **2004**, *46*, 427–441. [CrossRef]
42. Soltis, J. Passivity breakdown, pit initiation and propagation of pits in metallic materials—Review. *Corros. Sci.* **2015**, *90*, 115–221. [CrossRef]
43. Liu, C.; Chu, P.K.; Lin, G.; Yang, D. Effects of Ti/TiN multilayer on corrosion resistance of nickel–titanium orthodontic brackets in artificial saliva. *Corro. Sci.* **2007**, *49*, 3783–3796. [CrossRef]
44. Dong, H.; Sun, Y.; Bell, T. Enhanced corrosion resistance of duplex coatings. *Surf. Coat. Technol.* **1997**, *90*, 91–101. [CrossRef]
45. Wu, Y.; Zhou, S.; Zhao, W.; Ouyang, L. Comparative corrosion resistance properties between (Cu, Ce)-DLC and Ti co-doped (Cu, Ce)/Ti-DLC films prepared via magnetron sputtering method. *Chem. Phys. Lett.* **2018**, *705*, 50–58. [CrossRef]

MDPI
St. Alban-Anlage 66
4052 Basel
Switzerland
www.mdpi.com

Coatings Editorial Office
E-mail: coatings@mdpi.com
www.mdpi.com/journal/coatings

Disclaimer/Publisher's Note: The statements, opinions and data contained in all publications are solely those of the individual author(s) and contributor(s) and not of MDPI and/or the editor(s). MDPI and/or the editor(s) disclaim responsibility for any injury to people or property resulting from any ideas, methods, instructions or products referred to in the content.

www.ingramcontent.com/pod-product-compliance
Lightning Source LLC
LaVergne TN
LVHW070634100526
838202LV00012B/800